# 50 RIGGING CALLS

Chris Higgs and
Cristiano Giavedoni

ENTERTAINMENT
TECHNOLOGY PRESS

Application & Techniques
Series

# 50 RIGGING CALLS

Chris Higgs and Cristiano Giavedoni

etpress

Entertainment Technology Press

# 50 Rigging Calls



First published June 2014
Entertainment Technology Press Ltd
The Studio, High Green, Great Shelford, Cambridge CB22 5EG
Internet: www.etnow.com

ISBN 978 1 904031 75 8

A title within the
Entertainment Technology Press Application & Techniques Series
Series editor: John Offord



CODE / RC001-0614

# CONTENTS

# FOREWORD

This book contains a collection of 50 articles written between 2005 and 2014 and first published in *Lighting & Sound International* magazine, published by the Professional Lighting and Sound Association. I am grateful to editor Lee Baldock for giving me a platform from which to speak every two months.

I have avoided making any changes to the original articles except where it could be confusing to repeat the original, such as information about the National Rigging Certificate before it was launched in 2007. I have also edited references to what used to be called 'suspension trauma' because the 2009 HSE review should be regarded as the authoritative guidance on suspension induced orthostasis and syncope – serious stuff, and I'm not a doctor.

Otherwise, this is a collection of personal observations over what I believe to be a decade of significant change in the rigging community, generally for the better.

The collection is illustrated by my colleague Cristiano Giavedoni. The cartoon character Mario, figurehead for the EU rigging community, arrived in the UK at the first Rigging Conference in 2010 and as I predicted then, we will be seeing a lot more of him in years to come.

*Chris Higgs*
*June 2014*

WORK AT HEIGTH
REGULATION

# Rigging Call #1

*February 2005*

### Roundslings and UV

There have been enquires recently about the effect of UV on roundslings. There is plenty of evidence that UV degrades polyester, and emerging study suggests even the UV content of artificial light has an effect. The HSE have recently published guidance (INDG 367) because of the effect UV can have on harnesses and lanyards. This guidance is predominantly aimed at users such as contractors in the construction industry who by and large employ the equipment outdoors in harsh conditions. Don't confuse polyester webbing in PPE (harnesses, lanyards, etc), certain ropes or webbing slings with roundslings. These are affected over time; they are not usually 'sheathed' like a roundsling. Five years is generally considered a life expectancy by many PPE manufacturers. Polyester fibres exposed to UV radiation will be considerably weakened over time.

Many manufacturers use dyes and coatings to help protect their products and issue guidance on the expected lifetime of them. Roundslings used indoors in our industry are highly unlikely to have the inner load-bearing fibres exposed to UV during their lifetime. The outer sheath protects them from UV as well as cutting and abrasion; if the sheath is damaged and the inner yarn exposed, the sling should be removed from service and destroyed in any case. It is possible that slings left outdoors in strong sunlight for long periods of time (years) would be affected. Under normal circumstances in the UK, roundslings will not be affected. Indoors in most venues even artificial light is unlikely to be a problem, although lighting left on for long periods may eventually have an effect. If you use roundslings in an application where you feel there may be a UV exposure problem, speak to the manufacturer and get advice from your insurance company.

If you have any doubts, speak to the manufacturers of your slings and in the meantime, use a steel instead.

### Work at height PPE

Singing Rock harnesses from Czechoslovakia: I particularly liked the 'Profi Worker' style which is an EN361 harness (can be used for fall arrest). Amongst other nifty design features it has a rear extension lanyard allowing users to easily attach inertia reel block hooks to the dorsal attachment without pulling any muscles. Anything that helps people use the PPE provided is a good thing. Another neat product, the 'Frog' was shown at PLASA 2004 by Rope Assemblies who import the product from Italy. This quick release connector can be used one-handed to connect to a variety of anchor points up to 11mm diameter and is intended to be an alternative to karabiners. The device conforms to EN362, and having a breaking load of 22kN, presumably if used with an energy absorber and suitable anchors, it may be suitable for use with some fall arrest equipment in some situations. For work positioning it could be invaluable. A word of caution, though – even though it will fit, don't ever attach it to a diagonal brace in a lightweight truss!

Petzl have several neat new products, and their long awaited ASAP will be with us soon. The ASAP is an automatic device that follows the worker up and down vertical or inclined fall arrest ropes. Should the worker fall or even abseil too fast, the device locks and will arrest the fall. This is exactly what the industry needs to replace existing rope 'chucks' that bite the rope and are prone to slipping. The ASAP uses a revolving toothed wheel to grip the rope and appears virtually foolproof to use. Some training in use is essential as it is for all Category 3 PPE. Rigging Call believes it surprisingly inexpensive for what it can do.

The other item of interest is the Freino karabiner. This is a twistlock offset D karabiner to EN362, but it has a cunning extra. There is a hook with a spring latch forged into the backbone which is specifically designed to allow an abseil line to be trained over it from a descender, creating extra friction for use when lowering heavy loads in a rescue situation, or in a situation where extra friction is required. The traditional methods of taking the line back through another karabiner are gone. If you use a Stop or I'D, the Freino will be very useful. There are new helmets and accessories, more of which in future.

### Standards News

The Work at Height Regulations are supposed to be coming into force in early 2005. The date has been put off before, but HSE has said that early 2005 is the most likely date. There has been a lot of press and discussion about the Regulations, not least PLASA's seminars during the year, if you have not yet started to gear up your systems of work and equipment inventory, now may be a good time!

The other significant news is that British Standard 7906: Part 1 will be published in the New Year. It is snappily titled: 'Use of lifting equipment for performance, broadcast and similar applications: Code of Practice for installation, use and removal of above stage equipment', excluding trusses and towers – (these are covered by BS 7906-2:2000, Lifting equipment for performance, broadcast and similar applications – Part 2: Code of practice for use of aluminium and steel trusses and towers). The title says it all. It should be useful in all areas of the industry, including exhibition and conference production as well as theatre and arena rigging. There is guidance on safety factors, planning and equipment specification that follows on from BS7905: Part 1 which deals with the design and manufacture of the equipment we use. If you are a rigger, freelance or in a company, you will need access to this Standard.

Finally, as reported in PLASA Standards News in July BSEN13411 is a new standard which removes any doubt over how many dogs you need in a wire rope eye, what type to use and how they should be fitted. The Standard which is in five parts covering thimbles, dogs, splicing, wedge sockets, ferrules and sockets.

ALLOWED
NOT ALLOWED

# Rigging Call #2

*April 2005*

## STAC (Special Theatrical Alloy Chain)

Entertainment industry rigging often requires a safe way of adjusting rigging at height. Fitting a standard alloy bow shackle to a link of an appropriate long link chain makes for a quick, simple and safe system of work. There have been a number of reports of insurance company engineers advising clients that long link chain should not be used in lifting. This has resulted in venues withdrawing the chains from service in order to comply with their terms of insurance.

There are several common uses for long link chain in the entertainment industry. In theatres, light duty long link chain is often used at the bar end of a counterweight flying line for adjusting horizontal trim and heavier chain as centre bridle legs on three-line winch bars. These were often end welded and of wrought iron, not Grade 80 steel. Best practice demands better methods these days, but there are still plenty of long link chains around, particularly as 'barrel chains' used to link counterweight barrels or rig bars from counterweight barrels. In 'arena' rigging, long link chain is used as a method of adjusting the length of suspensions and in secondary suspensions to reduce slack (a Good Thing).

Often known generically as 'deck chain', Special Theatrical Alloy Chain (STAC), originally intended for cargo lashing, is made from Grade 80 alloy steel and meets the U.S. National Association of Chain Manufacturers specification for overhead lifting. The manufacturers, Columbus McKinnon, consider it suitable for use in the entertainment industry and provide specific instructions for safe use. It is intended to be used for entertainment rigging in a way similar to its intended use, i.e. in straight line tension, never as a sling. There are similar chains, such as Trawlex which may also be suitable, but also end-welded long link chains having no SWL and (if you look) the manufacturer's declaration shows that they are not suitable for lifting.

The Supply of Machinery (Safety) Regulations 1992* implements the EC Machinery Directive and only allows short link welded chain to be used for lifting (Regulation 2.2, and Schedule 3, at 4.1.2.5 (b)). Shorter

* The Supply of Machinery (Safety) Regulations was updated in 2008.

and thicker proportioned links offer better resistance to bending in chain slings. Sling chain may be twisted, wrapped around objects or have loads landed on it. In these situations the links in long link chain will not transfer load efficiently and the links may be overstressed by bending or twisting along their long axes. Incorrect fitting of shackles or hooks can spread links open. None of these circumstances should occur in the specific use the entertainment industry makes of long link chains. Using the recommended factor of safety of 8:1 (q.v. *Technical Standards for Places of Entertainment*, BS 7905:1 and BS 7906:1) STAC has a SWL of over 2.5 tonnes, more than adequate for the majority of entertainment industry rigging tasks.

Safe working at height includes reducing the time spent at height and using loose parts at height; using STAC helps satisfy both these requirements. Most rigging equipment is easy to use incorrectly and create potential hazards. In fact even when used *incorrectly*, STAC is probably safer than other methods. There are alternatives, of course; chain slings with shortening clutches are used widely and do conform to the Regulations.

This rigging anorak reckons that provided a risk assessment is carried out, the manufacturer's instructions are followed and STAC is treated as a lifting accessory (as defined in LOLER) their use is as safe as many other aspects of entertainment rigging *provided the links are subject only to axial loads*. However, a quick answer is unlikely and our efforts may need to be directed at Europe rather than London.

## Equipment news

LeMark has launched a disposable hazard tape for alerting people to your work overhead. The wide polythene tape is black on yellow and has the advantage of a printed caution message and symbols. This not only defines a hazardous area but also advises of the hazard the tape alerts you to. Crossing over such an obvious line, one would have to shoulder a large part of the responsibility for subsequent injury or damage to property. (Not that it means you can continue to work above; you still have influence on other people's safety).

Another Petzl product! The development of the Ecrin Best, the Vertex Best (A16) is here. It is very light and easy to wear with good visibility and a single rear adjusting wheel. If you want to, you can fit face and hearing protection into slots provided for the purpose. The helmet conforms to EN397, the industrial standard for head protection plus the benefit of a chinstrap like a mountaineering helmet. An industrial helmet strap should allow your head to slip out before you get trapped or strangled by the strap. If you're going to wear a helmet, why not make sure it won't fall off when you do? You see, there are clever people that design all this kit, so we get home in one piece. Find out what to use and learn how to use it.

### Training/education/NRC news

The Americans have announced the rigging experts working on the ESTA Rigging Certification Program. Roy Bickel, Rocky Paulson, Bill Sapsis and Harry Donovan are amongst them. I look forward to seeing what they come up with.

### Legislation/Standards

The British Standard BS 7906 part I mentioned before is now available. It is worth getting for page 56 alone. As reported in *LSI* magazine, there is a draft Dutch NEN Standard concerned with safety factors in equipment for lifting over people. It's worth a read.

PERIODIC INSPECTION
SCHEME OF EXAMINATION

# Rigging Call #3

*June 2005*

### Testing

Talking about "testing" lifting equipment often seems to confuse. It has been a legal requirement for a long time, but what does testing mean? It seems many people don't know that much of the equipment they use on a daily basis has never been tested with a load, as they suppose it has. Technically, equipment must either be new or have a report of thorough examination signed by a competent person before it is used, sometimes both; and in many ways that is all that we need to know.

So what constitutes this "testing"? The more complex the equipment, and the greater the risks in using the equipment is, the more exacting the test. A good deal of information can be gained from visual inspection. If the person conducting the test is competent, they should be able to determine the sort of life the equipment has already had and might be expected to have in the future.

The competence required may be defined as having sufficient knowledge and experience to determine if the equipment being tested is free from defects likely to affect safety.

The experience of the person carrying out the test is as important as the test. Many theatre flymen would have no idea how to test a chain motor without training and information; by the same token, many hoist engineers would not be competent to inspect a counterweight set without understanding how they are used. Other lifting equipment may need to be thoroughly examined by several competent people – electrical engineers and mechanical engineers, for example. The Lifting Operations and Lifting Equipment Regulations (LOLER) is clear that the nature of the test or thorough examination should be determined by a competent person. The Approved Code of Practice (ACoP) for LOLER does not require independent (third party) testing, as is often assumed. Many users of lifting equipment already undertake their own inspections of lifting accessories. The inspection of hoists, trusses, chain slings and more complex equipment may need to be outsourced if the necessary competence is not available in house. Check with your insurers, they may also require certain criteria to be met. The competent person appointed by the employer must have 'authority and independence to ensure that examinations are properly carried out and that the necessary recommendations arising from them are made without fear or favour'.

Many tests are visual and follow simple procedures to ensure all aspects of the equipment are covered. A shackle, for example, would be checked for obvious deformation, corrosion and poor operation. Dimensions and other features can be checked against manufacturer's information to support decisions. However, the cost of a shackle compared to the cost of the consequences of it failing makes the decision to replace it a no-brainer.

A hoist requires a more detailed look. It will need identifying and its 'service history' checked, but functional tests of things like brakes, limit switches and overload devices are necessary. In the case of a hoist brake or overload device this will mean a load test. This should not be so severe as to damage the hoist, and the hoist is designed to be tested with load; a competent person will know how.

A truss requires a visual test, relatively easy, but time consuming. Every member should be checked for true, gouges and dents. Each weld needs checking but if the truss is painted it is sometimes impossible even with specialist knowledge. In summary, if you think you are competent to inspect your lifting equipment in compliance with LOLER, you are probably already doing so. As the old saying goes, if you need to ask if you are competent, you probably aren't.

## Training and education news

ESTA's Entertainment Technician Certification Program (ETCP) 'Certified rigger program' (sic) is being launched later this year.

There are two certificates, 'theatre rigger' and 'arena rigger'. Each examination costs $600; sitting both in the same year attracts a $200 discount. Members of IATSE, ESTA and other trade bodies benefit from a discount.

The panel of ETCP experts worked with psychometricians to produce the computer-based, three-hour, 150-question multiple choice papers drawing on the ability to recall information, apply principles and assess the suitability of given solutions to problems. There is no practical examination, but there is a 30 point entry requirement, points being derived from work experience or education or a combination of both.

Based solely on work experience, a candidate would need proof of about three years' work (3000 hours) equal to about 20 hours rigging per week, or 18 months full time rigging employment or in apprenticeship. A postgraduate degree in an entertainment technology subject provides ten points, leaving about a year (2000 hours) of documented work to provide the rest. It seems a well thought-out scheme, and the UK rigging community will follow its progress with interest. The first examinations are due to be held at LDI in November. Congratulations to ESTA in being the first to deliver a dedicated rigging certificate for the entertainment industry.

## Legislation/Standards

As you may know, the postponed Work at Height Regulations will come into force on the 5th of April 2005, so by the time you read this they will finally be in force. Although the Regulations really only represent existing best practice, there are many areas where they will have an impact on rigging work. Rigging Call will be devoting the next edition to some extra information regarding how they may affect the entertainment industry.

## Kit

A neat bit of kit for a fiver: the "Ergodyne Squid" (squid?) is an elasticated lanyard for small tools up to 6kg. They come with a small 'do not use for climbing' karabiner you may want to replace, but otherwise seem a vast improvement on the curly type beloved by lampies. A cord lock secures the elastic around the tool if no hole is provided. It's a useful way of reducing the risk of tools falling when working at height.

DOUBLE BRAKES

# Rigging Call #4

*August 2005*

### Double brakes

When should you consider using hoists with two brakes, and why? Often it is an attempt to avoid the time and effort in using secondary suspensions ('safeties'). One could argue the likelihood of hoist failures is small, as far as we know. Well maintained hoists, used correctly with a sensible load share are unlikely to fail. However the consequences of failure may be catastrophic, particularly when people are beneath; safeguards need to be proportionate. Regulation 8 (1) (c) of the Approved Code of Practice for LOLER and guidance notes 234 and 235 may apply although this technically relates to people working under loads rather than when loads are suspended over people.

There are several ways to stop a load falling, typically with secondary suspensions or where the load needs to be able to move as part of the show, using load arrest devices. The provision of secondary suspensions also requires assessment to ensure that should a hoist fail, the load will not alter its position significantly or create dynamic loads that may themselves create additional, possibly greater risks.

*Technical Standards for Places of Entertainment* (ABTT) states secondary suspensions are not normally necessary in permanent installations because conditions and usage are more controlled. Independently anchored, almost taut secondary suspensions will stop a load falling; a 'motor bypass' cuts the hoist out of the load path. Accidents have often involved chain to hook connections; these would not necessarily have been prevented using a 'motor bypass'. Using shortening clutches on hoist chain creates extra risks and the two main manufacturers are not keen on this practice. There is nothing in law that requires the use of a 'second' brake in the UK. Although BS7906:1 suggests use of a second brake, this is not legislation.

A second brake may increase safety under certain conditions, but only protects against primary brake failure. Perhaps because it is offered by manufacturers, it seems to be highly regarded as a preferable specification, certainly providing a second brake is a commercial opportunity few manufacturers pass up. They are responding partly to German Regulation BGV C1 "Staging and production facilities for the entertainment industry" (1998). Although the Regulation does not specify double brakes, where hoists are permanently installed overhead in entertainment venues it calls up DIN 56925 (scenery hoists: safety and testing) and *in Germany*, BGVC1 hoist systems must be used. Bear in mind there are many other conditions a system must meet to comply with BGVC1; the position of the load limiting device in the load path is also specified. BVGC1 'does not preclude other at least equally safe solutions which may have been set out by other EU states' although it seems a fairly sensible set of requirements in the main. The ASM world-patented dynamic irreversible worm gear (shown at the ABTT Theatre Show this year) which prevents the load from slipping at all, even under extreme overload, is one such development. It is interesting to note that the distributor specifies the hoist as meeting full BGV C1 standard when used with an ASM Genesis control system – note that the hoist alone cannot meet the requirements of the standard. As in BS7906:1, the VPLT code of practice SR2.0 (though not a standard) for the 'provision and use of electric chain hoists in the event industry' contains requirements where hoists lift and/or suspend a load over people.

So what can we expect a second brake to do? It acts on the same shaft as the primary brake and guards against brake wear, incorrect adjustment, possible contamination or other brake lining problems on the primary brake. However, one well known manufacturer believes fitting a 'safety' hook may prevent more accidents than a second brake; two brakes won't prevent loads slipping out of hooks, a cause of many accidents. In summary, double brakes may be of benefit when dynamic loads cannot be safeguarded in another way or when using secondary suspensions are not practicable. The onus is firmly on the user in the particular application, something over which the manufacturer has no control.

The Work at Height Regulations came into force on the 5$^{th}$ April, as expected. PLASA have done a much better job of writing broad guidance than can be done here and I thoroughly recommend the guidance note to you. There is also a short HSE guidance leaflet (indg401) available from the HSE website which gives useful information. In the next few months, I am sure we will see 'sector specific' guidance we can use in addition to our own. As with LOLER, if you were doing it right before, you'll probably comply; there is little that changes existing good practice. However, read the information; you may discover you weren't using best practice to start with! I would draw your attention particularly to 'fragile surfaces' (and what those may be in the entertainment industry), specifications of 'suitable' platforms and the non-standard use of scaffolding. Rescue planning and formal training requirements may come as a surprise!

# Rigging Call #5

*October 2005*

## More on Working at Height

Walking around during the build-up of a well-known entertainment industry trade show this year showed that a great many people use Personal Protective Equipment (PPE) when working at height. Sadly, most of these folk were climbing on trusses. This may seem to be a good thing – under certain circumstances. The general principles of prevention in legislation state that PPE is the lowest tier in a hierarchy of control measures. For example, hard hats may be used to protect workers from falling objects in a work place, but only after other more effective methods are in place or are not possible – such as preventing access to the hazardous area, providing covered traffic routes and so on. The falling objects only affect the people who need to be there: they use the hard hats. Other people are excluded and therefore do not need to use PPE; this requires planning and supervision to ensure the method chosen is applied and adhered to.

In basic terms, the Work at Height Regulations require employers and the self-employed to ensure that the method exposing workers to least risk of falls is chosen for tasks that must be carried out at height, subject to the duration and nature of the task. You could argue that it feels safer scampering around a truss than being on tip-toe up a ladder, but that would be accepting the risk of serious injury, or worse, when safer options were available. In light of the potential severity of a fall from a height, any measures that need to be taken to guard against them may need to be proportionally stringent.

Using the falling object analogy again, the hazard needs to be controlled by designing the work to minimise the possibility of falling objects where possible, and if not to enclose the danger area and as a last resort to physically guard or at least delineate the danger area and clearly sign the hazard to those who need to enter.

I watched a chap sitting up on a lightweight triangular (apex down) truss structure, approximately 4.5 metres above the floor. He was wearing a basic industrial fall arrest harness with a two metre lanyard connected to one top chord of the truss he was sitting on. He had climbed up the truss tower and hooked up to the top chord of the horizontal truss. He climbed up past it and sat above it, ready to receive the moving light his colleagues were hauling up with a rope and pulley also rigged to his spot on the truss (!) Should our PPE'd friend slip and fall (easy on a sweaty afternoon) he would have fallen, possibly on top of the moving light but certainly onto the various flight cases and production detritus beneath.

Why? Because even if the energy absorber didn't deploy, the 2m lanyard, plus the distance from the harness attachment point to his feet would have allowed only about 500mm clearance; this before allowing for any deflection in the truss, the height taken up by the equipment and possibly poor fit of the harness allowing a further distance to be fallen. If the energy absorber did deploy even only 200-300mm, he would have hit the ground. I don't have space to discuss the force on the truss, possibly between 2.5kN - 6kN.

The thing that bothers me is that he thought he was safe because he was using 'safety' equipment. His manager regularly gives him a grief about using his harness, so it is understandable for him to think it important and necessary – so he complied. Why don't people using structural systems, usually practical people, get it? They certainly can't have much understanding of the equipment they use, and haven't read the instructions that a (Category 3 PPE) product bearing a CE mark must be provided with. The really interesting part is that in the middle of the stand in question were the Zarges they had used to rig everything else with…

As a point of interest, the maximum allowable extension of an energy absorber (using 'tear web' as the absorbing element) in Standards used for fall arrest is 1.75m. With a standard 2m fall arrest lanyard, this means the total deployed lanyard length could be 3.75m. To that you need to add the distance from the attachment point to the wearer's feet, maybe another 1.5m – meaning that the anchor point for the lanyard should allow at least 5.25m to fall into before taking into account the other factors such as deflection of the supporting structure, objects beneath the fall area, and so on. The further the distance fallen, the greater the force that needs to be absorbed, resulting in a greater extension of the lanyard. This is of particular importance when the fall may occur above the anchor point – although it is possible that the energy absorber may deploy fully in a short fall, it certainly could in a longer fall. (Another good reason to use a short lanyard.)

Another sighting was a lampie, with spanking new kit, lying face down on a truss focusing. His shiny new climber's helmet clipped to his harness, his lanyard neatly wrapped around his waist; around £300 worth of PPE – is it just a fashion statement? If he'd decided that there was little point attaching himself to the truss (agreed) then why not remove the grass skirt of lanyards, slings and other things people seem intent on snagging themselves with? The argument about attaching fall arrest to trusses is a moot point – there aren't many trusses that when loaded will sustain a dynamic 6kn point load on one chord. In the cases above, ground supported exhibition stand truss structures would almost certainly be pulled over or suffer some kind of progressive failure in such situations.

The industry needs to recognise that providing a suitable means of access for the work being carried out is a legal duty of the employer, even before the introduction of the Work at Height Regulations this year. It may cost more, it may take longer, but using appropriate access equipment is necessary; climbing on trusses is not generally a safe system of work. Sadly, a prosecution is probably the only thing that will bring this home to people. Using a set of Zarges to hang a few 500W Fresnels is probably entirely justifiable, as is their use in set construction, but climbing on structures not designed for human support is no longer an option, much less attaching a fall arrest device that may impose a significant force on the structure.

There were several stands that were using mobile access towers of various sizes, so it can be done. There was one contractor using a single width tower with a platform at over 3m with no outriggers. As a final observation – the house riggers were using work positioning lanyards in their access machines, peakless hard hats with chin straps at all times during the day and used 'banksmen' on the ground to act as sentries and a second pair of eyes. As sure as eggs are eggs, some will not have adhered to procedures as they should, but as a practical illustration of safe systems of work it demonstrated it is possible, and without vast expense or over implementation of regulations.

There is an assumption made by many suppliers that using PPE as a system of work is the preferred or only way to do the work. The principles of work at height are the same as in any area of risk: reduce risk by designing work to eliminate hazards, and where that isn't possible reducing exposure to hazards, dealing with the remaining hazards according to the same hierarchy. Many suppliers of 'height PPE' have built their businesses servicing the construction and engineering sectors where such techniques are practicable and justifiable.

The advice available to the entertainment industry from those suppliers may therefore assume factors with which the suppliers are familiar but our industry historically never consider; in fact, there often isn't a risk assessment in place, let alone method statements or company procedures. Yes, a competent person would consider rescue plans, but we are an industry that, by and large, still believes wearing a harness makes work 'safe'. We are not alone in this view, of course. Since we are also an industry dedicated to what we do, we perhaps ought to apply the same rigour to work at height as we do, say, to electrical safety. (Some may also add rigging to this list…) A safe system of work can be defined as "the integration of people, articles and substances in a suitable environment and workplace to produce and maintain an acceptable standard of safety. In this system, due consideration should be given to foreseeable emergencies and the provision of adequate facilities".

Therefore the provision of PPE for work at height – harnesses, lanyards, helmets and so on, is not the end of the story. Very often employers allow employees to select and purchase PPE themselves. This can lead to a lampie being given £200 from petty cash and letting him go to the nearest outdoor shop. Bringing their own climbing gear into work – even if self-employed may mean that equipment not meeting the standards of the PPE Regulations is being used in situations where compliance with standards is necessary. This might have serious repercussions in the event of an accident, practically and in terms of insurance cover. Even when such PPE is compliant, the employer should be able to check and approve the PPE for use in his business, rather than assume the user is an expert. Selection of PPE should take place after a risk assessment has been made and eliminate hazards at source wherever possible, controls being designed for remaining hazards and suitable PPE being selected for hazards that cannot be controlled in any other way.

NRC 2007
NRC

# Rigging Call #6

*December 2005*

## The National Rigging Certificate

You may have heard about the National Rigging Certificate. The idea was originally a joint scheme between Safe Working Ltd and Total Fabrications Ltd (Total Training) in 2003. The idea was to get all the rigging companies and players in the field to agree a common goal – the formation of one recognised certificate of competence to assist employers, venues and individuals to demonstrate competence in the field of entertainment rigging. The decision was taken by a show of hands in 2004 that PLASA should be the body to administer the scheme to ensure transparency and authority. Some time has passed and quite a few people have been very busy, notably Nicky Greet and Ron Bonner at PLASA. The scheme is well and truly in motion. In early October at the inaugural meeting of the advisory group to PLASA, we were tasked with defining rigging so the range of skills can be assessed in line with national occupational standards.

So the first task – what do we mean by rigging? It means different things to different people; if you work in theatre it is concerned with hanging pretty much anything your trade uses. If you work in television it means much the same thing, but the things that are 'rigged' are quite different and technicians use different techniques and equipment to do it. Film riggers may well use scaffolding, exhibition and conference riggers may use a range of equipment from other sectors within the entertainment industry.

One dictionary definition of 'rigging' is "to make or construct in haste or in a makeshift manner". This may not be how we want to be thought of! The essence of rigging is to position or move an item for production reasons using suspension or lifting apparatus, such that equipment used is in tension, but it may include supporting things from below, for example ground support systems. Difficult, isn't it? What do you think of as 'rigging'? Power cables, drapes, art or advertising material? Lighting, sound or AV equipment? A combination of all these things? The equipment used in the activity?

One of the primary differentiations is if the rigging is temporary or permanent. I would argue that a permanent installation of say, a flying system, is undeniably rigging (read on), but being permanent it may also be regarded as a work of engineering construction rather than essentially rigging.

I think, subconsciously, many of us think of rigging as being essentially temporary in nature; creating something where there was nothing, because it isn't necessary or viable for the purposes of the event to make the rigging permanent. If you do something permanent, the main contractor or building owner will use systems that ensure an adequate degree of competence and know-how are employed by all the contractors on a project. Traditionally, shows have been put on by enthusiasts, paid or not, and as such have never been held up to scrutiny in the way an engineering or construction contractor would be as part of a project. In the light of the compensation culture and the general maturing of the entertainment industry, those days are fast disappearing.

I contend that the easiest way to define rigging is "to support objects from above using lifting or suspension techniques". Anything used to support equipment from below is ground support, scaffolding or stage engineering. Using a ground support tower involves rigging elements (the hoist and slings, etc) but is basically a construction skill. Scaffolding to support a stage or scenery is not rigging. The activity exists in other areas in exactly the same way, although originally for other reasons and can be certified by other bodies. I agree absolutely that there is a need to recognise these skills and possibly to incorporate them into the skills base for the national certificate. 'Rigging' as far as the entertainment industry is concerned is therefore means of lifting or suspension of any production related equipment or effect, static or dynamic, or installing the equipment so used. This may include running power cables or draping off areas of venues or installing equipment for aerial effects or stunts. It may involve work at height and use of PPE, but it doesn't have to. The work may be in a theatre, concert or exhibition hall, arena, outdoor site or studio, but in all cases should relate directly to creative work, drama, dance, music or exposition.

Whilst we are thinking about standardisation, consider the formalisation of work at height training. The HSE Advisory Committee on Work at Height Training (ACWAHT) is concerned with just that. The committee has as its maxim "To collect and evaluate data and provide guidance on training to achieve the competence required by law for all those involved with work at height." The idea is to agree a syllabus that allows people who work at height to have a basic understanding of the law and how it relates to their work and workplace and to create a standard of competence. The committee has a broad representation from across most sectors of industry. The interests of PLASA, PSA and ABTT are represented on the committee. The level of awareness of work at height issues in the professional rigging industry in comparison with other sectors is apparently quite high, particularly with regard to access machinery and PPE for work at height. The theatre and corporate event market are lagging a little behind, but typically, the company or freelance rigger is at least the equal of his counterparts in say, the emergency services. That should give us some encouragement and help further the advances we have made in the last few years.

I was concerned for some time that we would be overtaken by general industry and become the poor relation. I think in hindsight that has not occurred, however, we still have some way to go in terms of recognising that we have to stand up to those seeking to cut corners for financial gain. Planning work to ensure safety should never be allowed to be subjugated by commercial interests.

So we have some considerable recent advances in development to be aware of. Those involved in rigging and suspension work above stages have two British Standards (BS 7905 and 7906) to use in ensuring safety and in the way the equipment is specified and used. We will soon have a national assessment and certification scheme (the PLASA National Rigging Certificate) to measure the competence of those working as riggers. Those working at height in the industry will soon have a British Standard (the ACWAHT syllabus incorporated into a British Standard) to use in ensuring they have the understanding and basic competence in matters relating to work at height.

Interesting times. We are in the situation that many of us have been worried about for years, but it seems that we may just be able (by the skin of our teeth) to 'tell them before they tell us'.

INSPECTION CALENDAR
= CHAIN HOIST INSPECTION
APRIL
1 2 3 4 5 6
7 8 9 10 11 12 13
14 15 16 17 18 19 20
21 22 23 24 25 26 27
TODAY
MARCH
9
RIG OR DIE

# Rigging Call #7

*January 2006*

### Lifting Equipment Inspection Intervals

After eight years, the six and 12 months intervals between statutory inspections specified by the Lifting Operations and Lifting Equipment Regulations 1998 (LOLER) seem to be generally understood. I find, particularly in theatres (places where traditionally there was only an annual insurance inspection of the permanently installed flying equipment), that 'rigging gear' is examined at least on an annual basis and the equipment is often of a reasonable standard compared with perhaps ten years ago.

LOLER breaks lifting equipment up into three distinct areas. Firstly, 'lifting equipment' – anything used for lifting loads including the anchorages to the supporting structure, e.g. a counterweight set. 'Lifting accessories' which is lifting equipment used to attach the load to the lifting machine, such as slings, shackles, beam-clamps and even trusses. LOLER states lifting equipment should be examined on an annual basis, lifting accessories and lifting equipment for lifting people should be examined at least every six months, i.e. more frequently owing to greater wear and tear and risk to safety respectively. (The third area, lifting equipment for lifting people, is another subject for another time, but covers MEWPs, rope access and positioning equipment, and other lifting equipment used for lifting people, e.g. the Wizard of Oz's hot air balloon basket…)

A point of interest, to immediately digress, is that an inspection carried out by an insurance company may not be a 'thorough examination' as required by LOLER. Don't forget, the onus is on the employer to provide safe work equipment and as such the insurance company is not involved – they are typically assessing their risk, not yours. The insurance company should be consulted to establish this.

Meanwhile, back to the plot. The interval between inspections is an issue that can cause confusion. LOLER requires lifting equipment to be thoroughly examined by a competent person and to be carried out often enough to ensure the equipment remains safe to use in the circumstances encountered in the employer's activity. That is to say, the item must be inspected often enough to establish and maintain its integrity by someone who understands the item, material(s) and manufacture, the way the item is used (in the entertainment industry) and the consequences of any deterioration likely to take place.

What is not generally appreciated is that LOLER also allows a formal 'written scheme of inspection' to be used **instead** of a rigid six or 12 month regime; the ACoP states that if a written scheme is not used it assumed the 'traditional' six and 12 monthly intervals will be used. This may be a more cost effective and reasonable method for some. It is important to recognise the need for expert help in drawing up the scheme in certain areas – rigging and lifting equipment companies are a good initial source of advice. Here are a few things to consider in starting drawing up an inspection scheme.

- How often is the item used?
- Are some parts likely to be more worn than others?
- Can any of the equipment deteriorate in storage if it is not used? (e.g. natural fibres, wire rope, chains or fittings stored in damp conditions)
- What expected life do certain items have? (e.g. steel wire drifts – don't expect them to survive long when used in the traditional manner on flying bars.)
- Changes of use or changes in risk that should require a review of the inspection scheme.
- Have you any recorded data to support your view, for example from inspection or maintenance logs?
- Identify items needing simple visual inspection, those needing some training to examine, or where the necessary competence to enable thorough examination in house would be lacking.
- Identify the required level of inspection – simply visual examination, dismantling and visual inspection, with or without measurements, NDT (for example dye penetrant weld tests) operational tests, or load tests.
- Are competent personnel available (trained and experienced) in-house and are appropriate facilities (e.g. time, workshop, test instruments, reference material) to carry out a thorough examination on a specific item? This will vary from item to item.
- Where 'rejection criteria' information may be found (amount of wear, etc – usually from the manufacturer).

- A risk assessment should consider the environment in which equipment is used and the number and type of loads lifted or supported. Take into account the degree of risk associated with each item or part of an item when used overhead.

The competence required to assess a wide range of equipment to develop a scheme of inspection is probably the reason people find it easier to opt for the six or 12 month maximum periods, regardless of situation. Often it is simply 'having someone else do the paperwork' or the desire to use a third party (Note: The employer is still likely to be liable ultimately).

Make an inventory of all the equipment that needs statutory examination. Assess the complexity of the equipment and decide which items need to be contracted out, and which can be reliably examined by your own staff.

Certain items might be visually inspected on a weekly basis, defects being detected by staff and if you have enough kit, quarantined for subsequent inspection by an outside body at less frequent intervals. My guess is that roundslings and larger shackles could often be inspected in-house. The discard criteria are simple; the cost of replacement in comparison to the consequence of failure is small. Items such as barrels, trusses and so on could be inspected in house with support from suppliers, manufacturers, and with some training – so might certain types of sling. Chain hoists, winches or complex mechanical items will usually be a matter for a specialist company.

If one were to interpret LOLER literally, and provided equipment is of good standard, one might argue some lifting accessories in a theatre environment are not exposed to conditions likely to cause deterioration and that inspection wasn't needed at all! A scheme of inspection may be justified instead. Items used only for a small part of the year or only for a specific production would be particularly suited to this risk based approach.

The principal risk in the entertainment industry is that the equipment supports loads overhead; examinations must be made properly, but using rigid examination intervals may be unnecessary.

The use of a formal written inspection scheme is worth considering, but you need to take expert advice from a wide range of sources including manufacturers, designers, insurers, users and maintenance staff. Worth a thought?

# Rigging Call #8

*April 2006*

### Loads on Structures

The law requires equipment used for lifting to be strong enough and stable enough for the foreseeable forces to be imposed on it. This includes the attachments to supporting structures. The information on strength (lifting capacity) should be available to a user in an understandable form. This means you have to know how the load will be applied, and where to look for the information you need. It also means you have to know the load to be applied at a given point – not always easy.

If you can't find the necessary load information, you have only two choices: either take the decision based on your experience and judgement, or don't hang the load at all. Typically, the view seems to be that the decision has been taken already: "I have to hang the lights there". Do you? Naturally, you are trying to do your job. That is where the lights need to be. Is the need for ideal lighting angles more important than risking a potential structural failure during an event? In our industry such collapses are comparatively rare, but when it happens it can be catastrophic.

If all goes well, you're everyone's best mate, the most conscientious rigger in the world. Should something collapse, you'll find no-one else thought you should have hung it there either. While you (the 'competent person') are happy to take that decision, the rest of the production will be happy. That's your call, being the rigger. The pressures you feel are that the rest of the production are relying on you. You are understandably concerned about the risk to your reputation and future employability, and potentially your liberty if it all goes pear shaped (I'm trying to use the correct legal terminology).

Your options are to rig or not to rig? If it's patently obvious that the load is but a fraction of the capacity of the structure you are contemplating rigging from, you would probably continue. If it isn't that obvious, you need to discover what the allowable load is, or use an alternative method of support.

The traditional alternative to overhead rigging is a ground supported structure. How often has a touring production been prevented from hanging the show and opted for ground support? Has the same consideration been given to the floor as to the overhead structure? The perception, even by the venue, that the floor will be a 'safer' option may be flawed (guffaw) even if it is a concrete slab. Loads need to be considered.

Four-point loads on a suspended floor could be far more onerous that optimising the capacity of overhead structures. Spreading temporary loads is generally easier to achieve from above than below. Very often the available positions for towers or supports are limited because of sightlines, seating or access requirements. This may result in concentrations of load over small areas, typically less than a metre square, often approaching half that.

Allowable floor loading in buildings is often 5kN/m$^2$ (500kg/m$^2$). Stages often have a higher figure of 7.5kN/m$^2$, but what about when it's not being used on a stage? Consequently, a standard one-tonne ground support tower (well, that's what it's designed for) which will actually easily lift as much as 2.5 tonnes before you know it, could significantly overload the floor area it occupies, even before considering any other floor loads in the vicinity (audience access routes, other stored equipment, an evacuation or false alarm…).

With regard to suspension, think of the number of hotels that provide rigging points in their conference suites. The information (if available or considered) on the capacity of their rigging points is often given as a point load or load per metre. You have that information to use in calculating the amount of equipment you can hang, taking into account such things as the dynamic load created by motors which can significantly increase load values. Is the information given for a dynamic load? Not often.

Recently built 'conference' venues often have a system of hanging points or channels to provide low profile, maximum headroom rigging for lighting. These usually negate the use of trusses and chain motors; the load is well spread on multiple supports and lightweight components. This allows a reasonable payload – lots of lights – to be rigged; the trade off is the ball-ache (a technical term) of having to rig every luminaire, though neat, individually from access equipment, not to rig at ground level and push a button.

The other problem is that systems are often 'invisibly' supported. I have seen chain motors rigged to single eyebolts in a channel and 'towing' at an angle, twisting the channel significantly. Not a safe method, it would have been designed for static vertical loads and rated accordingly.

What if the allowable load is woefully inadequate for the proposed lighting design?

Very often the allowable load is exceeded by a large factor. In this case, were there an accident, the employer (typically the producer or production manager) may be held responsible for allowing the work to take place. Small, and some not so small venues seldom appoint a 'house' rigging contractor, presumably on grounds of cost, but such an arrangement could resolve many of these issues. Applied loads are typically pre-notified and the contractor has intimate knowledge of the venue and the ways loads can be applied.

Ballroom facilities often used chandelier hanging points for the rigging points for trusses for events. In the past these often had no loading information at all – the only data available was usually an optimistic estimate of the weight of the chandelier.

Even if this estimate was reliable, the practice was a bit dodgy (to use another technical term), since the chain motor that replaced the chandelier was likely to able to lift 1000kg (without monitoring) rather than the 500kg or so the hanging point was normally used to support. Moreover, a typical hoist lifting speed of four metres per minute creates an initial 'surge' load roughly equal to 125% of the load being lifted. Potentially 12.5kN on a chandelier fixing.

Interestingly, the Work at Height Regulations would now push architects and designers towards systems that allow as much work as possible to be done at ground level. More recently designed facilities often provide a reasonably rated eyebolt every six metres or so which means lighting can be rigged at ground level on trusses. Of course, the eyebolts are often no more than four metres from the floor, but you can't have everything. Architects, please note.

The fact remains that the law requires 'the employer', to ensure that the person responsible for the rigging knows the loads to be applied, the allowable load on the structure is known and the two are checked to be compatible by a competent person. It is incorrect to assume that because the lighting is relatively well spread out and the bar or truss is picked up in, say, four places along its length, the total weight is divided by the number of supports. Riggers should know that a uniformly loaded bar or truss that is supported by three or four points will need to be checked to ensure the centre point(s) do not exceed the allowable loads. In the case of chain motors, it is quite easy to change loads dramatically by simply 'bumping' the motors up or down, and consequently the actual point loads may be very different from the intended loads.

Without using a load-cell to measure each point load, or unless there is a very healthy margin allowed, the loads will almost certainly exceed the allowable load on the suspension point unless the rigger is extremely careful. Thankfully, the SWL of the rigging point provided should have an allowance for this built in, but legally the SWL should not be knowingly exceeded. The user is unlikely to know what safety factor was used in the design. A good estimate of the load is something every production should be able to produce; asking for the loading information from the venue is equally straightforward. Since British and European standards typically quote the safety factor to be used when lifting overhead is typically double that used for normal lifting, the rigger should be looking to maintain an 8 or even 10:1 factor in such situations.

As the experts, we should know that getting such information from the venue may well take weeks, and as such, we need to allow as much time as possible to get the information to work with.

Being pressurised at the eleventh hour is not acceptable in such a high risk decision, as a venue, a contractor or as a client.

RI
O
DI

# Rigging Call #9

*May 2006*

### Work at height update

It is now a year since the 'Work at Height Regs' came into force. Here are three popular work at height issues that seem to keep cropping up when people are thinking about safe systems of work.

### Tallescopes, Towers and Ladders

The Tallescope argument limps on with no apparent end in sight. Until either a ruling on their use is made or a viable alternative presents itself, it seems unlikely that a solution will be found. The bottom line is that they were not designed to be used in the way theatres use them, and the manufacturer's instructions do not allow for such 'mobile' use. If one disregards manufacturer's instructions, it seems the authorities may be minded to treat that as a breach of regulations. There are plenty of alternatives to them, a plethora of machines to suit all pockets. Their disadvantages are their weight, size and ability to be used on slopes.

The alternatives to Tallescopes where they are most useful, on theatre stages, are few – a mobile alloy scaffold tower is probably the most practical option. The footprint of a 2m narrow width (typically .75m) tower is little different to that of a Tallescope excluding braces. You can reach maybe four to six luminaires without overreaching from the platform but you can't move a tower with someone on it, either. So you still need to ascend and descend often. Other disadvantages include the time taken to build, strike it or change the platform level. At 7m, a tower may feel less stable than a Tallescope, too, even with an outrigger in each corner. Ensuring that the most appropriate means of access is used is a legal requirement and this may mean that a larger range of equipment is needed to comply with that requirement. This may mean less use of the Tallescope, therefore statistically a lower risk than at present.

The confusion over using ladders has hopefully subsided. You can use ladders, but only where use of a safer means of access is not justified. Light duty work for a short period from a ladder is acceptable, but users need to be using them correctly. There is a clear requirement for ladder users to be trained to know when a ladder is appropriate for a task, how to use them safely, and that employers ensure ladders are of good quality and well maintained.

## PPE

Using twin energy absorbing fall arrest lanyards incorrectly has been highlighted. The force on the user is normally limited to 6kN by the action of the energy absorber in the lanyard. Hooking the 'spare' leg of a twin lanyard to other attachment points (often those provided for work positioning) on a user's harness means that should the energy absorber deploy during a fall, lengthening in the process, the energy absorber is effectively bypassed. The remaining force generated by the fall is transferred via the 'parked' leg to the harness attachment point. A test I saw resulted in a 16kN force at the 'hip' attachment point, potentially fatal if the remaining force (now unchecked by the energy absorber) is too great for the lanyard, harness, anchor or the user.

Using lanyards or inertia reel devices for fall arrest on lighting trusses is still an issue. Both are regularly seen used with trusses that are already holding their design load. Lanyards of even 1m length can generate close to the 6kN mentioned above if the user falls from above the anchor point. 6kN applied as a point load on a truss that perhaps only has a 600kg allowable centre point load? To hang half a ton of lights and then apply a dynamic load of up to 6kN when someone falls cannot be a safe system of work.

Many inertia reel devices require the user to anchor them to a structure that can sustain 15kN. They should also be rigged so the device is always above the user. This is to ensure the device can act in the way it was designed and lock after the user has travelled only a very short distance. If the user falls from above the device, the force generated may exceed the capability of the device.

When used to protect someone climbing a wire rope ladder to a truss, the period of greatest risk is often getting back on the wire rope ladder to descend. In this case, if the device is rigged to the underside of the truss (which they often are) the user will fall past the device, creating a force which the device cannot arrest.

It is worth remembering that typical 'industrial' 2m lanyards may need as much as 7m of clear space for a user to fall into. The use of horizontal safety lines is often seen as the obvious way to protect people moving along trusses. The strength of a proprietary system is not in question, that is what they are designed to do – provided you know what anchor strength and span is specified, the number of users it was designed for and the clearance required if it is fallen on. The specified anchor strength is often 15kN or more, something few lighting trusses can sustain especially when already loaded with lights. When a 9m clearance is required, they start to look less like a safe system of work in such applications.

**At this point in the original article there was a short paragraph about how to treat what we then referred to as 'suspension trauma'. I've removed it from this document because this was a subject later researched by the HSE. As a result the notion of trauma was questioned and the advice was then, and still is now, to treat an unconscious casualty who has been hanging in a harness for a long period using the normal first aiders' ABC protocol (airway, breathing, circulation). This was updated a year later – see the July 2007 Rigging Call #15 and the HSE guidance issued in 2009. This is still available on the HSE website under 'Falls'.**

In closing, a quote from the Work at Height Regulations. Regulation 11 has this to say (more or less) about 'Danger Areas'.

*Every employer shall ensure that where a workplace contains an area in which, owing to the nature of the work, there is a risk of any person at work falling a distance, or being struck by a falling object which is liable to cause personal injury, the workplace is, so far as is reasonably practicable, equipped with devices preventing unauthorised persons from entering such area; and such area is clearly indicated.*

Just so you know…

# Rigging Call #10

*August 2006*

## Rigging Definitions

So what is rigging, anyway? From the entertainment industry point of view each sector has evolved its own meaning. In television, for example, rigging has more to do with the activity of hanging the lighting equipment itself because studios, like theatres, usually have an installed suspension system of some sort that allows technicians to hang luminaires directly, not very often needing to worry about the inherent strength of the structure from which the suspension system ultimately hangs. In outside broadcasts, rigging may have more to do with building temporary platforms, working with lighting stands and cable management than using lifting equipment. Theatres usually have a suspension facility of some sort and though Sod's law always applies, there is generally a bar fairly close to where you want it.

In both cases, it could be argued that the rigging can be undertaken by the electrics or scenic department; the systems are managed by technicians ('flymen' in theatres, who are not of necessity riggers), stage or scenic staff. Lastly, there are venues where no suspension facilities exist. Every production brings in everything it needs – rigging and now riggers are required.

It might be useful to consider some of the styles of rigging and the terms used to describe them now the trade is being (reluctantly?) accepted as an essential requirement in many cases. The days when lighting technicians were always considered responsible for rigging the motors and trusses for a production, or the stage crew for the flying of scenery and props for large scale productions have largely disappeared. More and more the activity is seen as a discrete process and a trade in its own right, very often rigging provision being made by the house and the riggers may be working for the venue not the production or promoter.

Production rigger/ing refers to the rigger/ing directly relating to the production (funnily enough) be it a touring concert or live event, a theatre show or a corporate event.

House rigger/ing is a term used to describe riggers and rigging who are generally supplied by the house to hang a production. This is usually to protect the venue's interests but also facilitates the loading in and out of a production since managements generally want to tour the minimum staff necessary. House riggers may be from a rigging contractor who services the venue, or the house may have full time riggers or employ them directly on a freelance basis.

Concert rigging is almost always a description relating to live music, usually popular music but it can include orchestral productions that tour arena and similar venues or festival and similar music productions. In the concert touring world, rigging is involved with providing or augmenting a suspension facility for the needs of the production. In such situations, rigging is the activity of installing and the equipment so used.

Event rigging is a hybrid term used for any live event, typically but not necessarily outdoors. This could range from a concert stage used for outside broadcast of a concert of other production, corporate or multimedia production. Rigging is often split into the rigging required to build an outdoor stage and then the hanging of one or more productions' technical requirements. Concert and event rigging may include stage building and/or scaffolding and other specialised rigging disciplines, from 'SkyCam' type camera rigging work to the opening ceremony productions at large sports events. The recent trends in supporting video walls and speaker arrays for stadium concerts have crossed over into other markets including sports events and ceremonies.

Exhibition rigging in larger halls is similar in approach to arena rigging; the equipment or objects being suspended are nothing to do with the rigging contractor as they might be for a production rigger. A rigging contractor will generally take a brief from a client, then liaise direct with the venue; this is similar to rigging a concert in many cases.

Museums, galleries and other attractions often require high quality rigging; the suspended objects could well be valuable, unique and even priceless, and be required to hang very precisely and unobtrusively.

Circus rigging is most closely allied to arena rigging in terms of technique; the main difference is that loads are almost always dynamic – and human.

Stunts and flying effects in film, TV and theatre use many methods found in other sectors and in some ways is closest to circus rigging since loads are likely to be human.

So that is at least eight separate rigging styles, before thinking about theatre, conference, sports and film rigging. Rigging is a skill that often goes unnoticed and more often unappreciated.

Like many 'manual' occupations, rigging can be perceived as simple. In fact it can be a hugely complex task in both practical and theoretical senses. Qualified engineers are an essential source of knowledge but require practical skills to realise their designs. Riggers are often able to 'imagineer' a solution to a production problem and then approach engineers to confirm or calculate forces and reactions in order to select appropriate equipment and structures from which to work. Using engineers is often necessary because a rigger is not (or often likely to be) a qualified engineer. The law requires sufficient competence in designing lifting operations that will often be in the province of engineers, not riggers.

Much of a rigger's work is impossible to record or describe. Often a job can be roughed out for commercial reasons or safety management, but the actual activity is unknown until the job is started. There is a skill in doing this, a kind of professional guesswork carried out in a defined time frame, with the resources estimated as being necessary to hand, but the job is usually something that has never been done before and it is unlikely to be possible to mock up or trial many projects.

A typical example of this is a recent project carried out by a well known rigging company where (real) aircraft were lifted, manipulated into 'action positions' and secured to a particularly unhelpful building structure. This involved some nifty rigging to install the hoists used to lift the planes, specialist airframe engineering and the design and fabrication of very special 'flying points' from which to lift the aircraft as well as structural engineers working to ensure the building wasn't being overstressed.

On top of that, museum culture is not rock and roll; the riggers may be considered aliens and the trust of a client to let riggers loose on valuable, if not irreplaceable objects has to be earned. The bona fides of riggers are often only recognised when work on site starts. Again, this is because of the unwritten methods and the invisible nature of rigging work. This is a common feature of all rigging work when an entertainment rigger has to fit in with other trades outside the entertainment industry. It would be fascinating to learn if this interface existed in the past in other areas of rigging – did ship's riggers find it easy working with naval architects? Do overhead power transmission riggers get on with pylon designers? The symbiosis of engineer and rigger is an important relationship, and one that must be developed to ensure that productions are safely and efficiently rigged – whatever type of rigging is being used.

04:15
AM
LEVEL 2
NRC
NRC
NRC
NRC
RC

# Rigging Call #11

*November 2006*

## The NRC

This month, the Rigging Call column is given over to update readers with progress on the National Rigging Certificate – probably the most significant development in entertainment rigging in the UK since the inversion of the chain hoist. The scheme is well under way, pilot schemes planned and the first assessments on track for the New Year. I'm advised there have been many obstacles to overcome and there will doubtless be more in the future, but the existence of a National Rigging Certificate is now a reality. For years, many of the rigging community have been aware of the lack of a formal method for riggers to demonstrate their competence, especially since there is no current legal requirement to meet a recognised standard. We all know that if the rigging industry doesn't do something itself, someone outside will – potentially something inappropriate to what entertainment industry riggers actually do. So here's an update which explains the scheme so far and why the PLASA team and the rigging community have been so busy.

It is now a year since PLASA agreed to develop the idea of a National Rigging Certificate. In that year a lot has happened and we are now in a position to give a more detailed update on what the scheme is about and what it means to a potential venue owner, employer or candidate. What is the overall aim of the scheme? In simple terms, it is to create an industry recognised qualification for 'Entertainment Rigging'. A skills card will be underpinned by a qualification gained through assessment against a set of written national occupational standards – in other words, a measurable test of an individual's skill against an agreed national benchmark. The standards once agreed are owned by government and incorporated into a qualification by PLASA. These standards are applied and used for qualifications across many other industries and will give entertainment rigging a recognised status as a profession.

On achieving the criteria in the standards, the candidate is awarded a certificate and a skills-based photo ID card for that level.

The card and qualification structure is configured to feature four levels:

Trainee Rigger
Level 2 – Rigger
Level 3 – Rigging Supervisor
Level 4 – Rigging Manager (Title TBC)

The key difference in the levels is one of responsibility and increased depth of knowledge. These levels equate to the National Qualifications Framework, the structure used for UK qualifications. One of the most fundamental aspects of the certificate is that it is an assessment scheme, not a training scheme. It is a standardised measure of an individual's competence. How someone acquires the skills to undertake the assessment is down to the individual – an approach we believe gives choice and flexibility and recognises the value of both formal training courses and work-based learning.

## Why do we need this scheme?

There is currently no 'National' formal mechanism in place to assess rigging skills, which makes it quite difficult for freelancers or companies offering rigging services to demonstrate to a potential employer or client their competence to carry out work. Concerns were also raised amongst the rigging sector that if the rigging industry didn't create a means to demonstrate competence, someone outside the industry would, and it could be a system wholly inappropriate to the work riggers do. In addition, the need for a Certificate has been strengthened by the recently introduced Working at Height Regulations, where there is a specific requirement that 'all persons working at height are competent to do so'. Whilst the majority of riggers working in the industry today are competent, this provides a mechanism to prove it. The scheme will also provide acknowledgement of 'entertainment rigging' as a profession in its own right and provide clear career choices for new entrants.

## Who agreed the standards?

The strength of the scheme is that the benchmarking for rigging skills has been set by members of the rigging industry itself through the Rigging Advisory Group, which features representatives from companies such as Nippy Industries, Outback Rigging, Summit Steel, Star Rigging,

Safe Working Ltd, Total Solutions, The Rigging Partnership, Unusual Rigging and Vertigo Rigging, together with venues such as Earls Court and key members of the freelance community. These companies and individuals represent a broad range of sectors including conference and corporate events, concert and touring, broadcast, theatre and live arts. This has ensured the development of an inclusive certificate applicable to anyone involved in carrying out the core skills and knowledge required for entertainment rigging work. As part of the process for National recognition, they are working with government bodies such as Creative and Cultural Skills, a sector skills council responsible for approving occupational standards through consultation with industry sectors.

### What is the qualification structure?

It is a modular framework made up of occupational standards (units), which require the candidate to demonstrate not only their technical rigging skills, but also how they communicate and work safely with others. The mandatory units cover communication, health and safety, and risk. Level 2 means the candidate is competent to work without direct supervision. At Level 3 there is an additional mandatory unit for supervisory and management skills respectively. Level 3 is the Rigging Supervisor's qualification. The units that have been selected cover the generic core skills required in rigging – regardless of the sector worked in. Meeting the standards required by all of these units will qualify the candidate to hold the National Rigging Certificate at the appropriate level. The framework also creates a possibility to develop the scheme so that individuals can have other skills assessed to assist in demonstrating their range of skills to potential employers.

### How will it work?

The plan is to approve assessment test centres around the UK to facilitate the assessment.

There is a three-step quality check in place to ensure the integrity of the scheme is maintained. Firstly, the assessment is carried out by assessors who themselves are experienced entertainment riggers and have been assessed to the same rigging standards. They have also undertaken the A1 assessors' award – a nationally recognised assessor's qualification.

Assessors are monitored by Internal Verifiers at the Centre and finally External Verifiers from the awarding body ensure assessments are valid, accurate, current, sufficient and reliable. In short this provides integrity and transparency and keeps it in line with other national qualifications.

To begin with there will be two routes of entry. The 'existing rigger route' will be available for a set number of years, so working riggers can achieve the Level 3 qualification without needing to begin at Level 2. Once the existing rigger route closes everyone will start as a new entrant, meaning training and work experience precedes assessment at Level 2 and then candidates spend a period of time working at Level 2 gaining experience and knowledge whilst working towards assessment at Level 3, the Rigging Supervisor's qualification.

More information will be available on the PLASA National Rigging Certificate website.

STATIC vs DYNAMIC

# Rigging Call #12

*January 2007*

## Forces

In a recent conversation with a structural engineer, I commented that it sometimes seems expected that riggers are able to walk into a venue, look up and instantly decide if the roof is strong enough to hang a show. The structural engineer's reply was: "Well, a structural engineer can't, so how can you expect it of anyone else?"

This decision may seem easy; the beams may be huge, the show very light, but are the beams that size for some other reason? Admittedly, sometimes you don't need an engineer, you can take a number of decisions based on the information that should be available from the venue or that you've already researched. The golden rule remains, you need to know the weight of what you want to lift and how that relates to where and how you attach it to the supporting structure. The structure may appear to be substantial enough, but two factors exist. A) is it really strong enough? (that is to say, how close to the wind are we sailing?) and B) How would the rigger demonstrate that due diligence had been exercised if it is ever required to do so? "But the beams are really big," is often a response. Was the venue provided with those beams to hang shows from or are they needed to hold the roof up so the snow doesn't come in? Is the span equally large? What are the foundations like? Is one able to assess the capacity of the steelwork as an individual beam, never mind the capacity of the bigger structure?

Let's suppose the beams are massive, with short spans supporting a concrete slab. How does the rigger calculate the load the beams are already carrying, before subtracting this from the capacity of the beam configuration (if known)? Has the dynamic effect been considered? This is usually taken as an extra 25% of the load being lifted at a typical four or five metres per minute provided by electric chainhoists. The figure will be a lot higher using faster speeds; this dynamic effect is created every time a lift is stopped or started.

Another example: a FOH lighting bridge supported by a roof truss – perhaps a perfect position for an advance lighting truss or a line array to be hung. Let's assume it was designed to carry its own weight plus its design load of lighting equipment, and an allowance for a number of people in accordance with the standards for structural steelwork in buildings. Now you arrive and add two 5kN loads four metres either side of centre. Who said you could? Are you prepared to take that responsibility because it looks like it should be strong enough?

Consider another situation. You do a 'recce' at a venue and ask where the rigging points are. You're told: "everyone uses those eyebolts up there". You duly rig to those points in the possibly mistaken belief you are 'covered' because you have been advised by the venue. To be sure, and to comply with LOLER, you need the venue to provide you with documentary evidence of the points' load capacity to assure you they have been inspected before you hang anything. "Nobody does that!" you may say. Far-fetched? Too much hassle? You decide: the next person that asks may be acting for the prosecution ...

Experience in hotel and conference venues shows that although the installed system may have been designed specifically for hanging lighting, contractors servicing productions very often don't have the information they should. Hotels often have either permanently rigged bars for lighting or sound to be hung from. Some have hanging points such as eyebolts or a channel fixing system. In either case these will have been designed to a brief envisaged by someone who may or may not have an understanding of the needs of events or productions. Even more often, they are used by people who don't know what that brief was and hang what they want to hang in the manner they devise. To provide a useful attachment point, an eyebolt may have an 'SWL' significantly greater than the allowable load on the structure that supports it and be marked as such; users may take that marking as the allowable load.

For example, to hang a truss using a short roundsling with an appropriate shackle – say a two-tonne alloy bow shackle – you'll need a 20mm eye. To obtain that size an M16 eyebolt is often required. This will typically be marked 'SWL 0.8t'. The supporting structure is often unlikely to be able to support this load, particularly when fitted to a 'channel' system. The eyebolts and fittings fit, but they are totally reliant on the strength of the channel and what supports the channel. Users may research the channel system product and discover that the fitting has an allowable vertical load of 9kN (10kN is approximately a one-tonne force), and using an 800kg eyebolt surmise that they can hang a chain block for a lighting truss or line array, lifting perhaps 300kg. However, the allowable load could actually be as little as 75kg. What holds up the channel? A quick look above the ceiling may be interesting. It is not unusual to find long unsupported lengths of channel, fixings being lengths of M10 all thread into timber joists, 'Hollorib' wedge anchors and the like.

It is not unusual for alloy bars and trusses supporting lighting, sound and AV equipment to interconnect and lend support to each other; a complicated loading situation. Common sense dictates that the more supports used, the shorter the spans, thus the lower the individual loads are on the structure. The practical upshot is that there are perhaps three contractors all using the same structure, not always coordinating or even being on site at the same time. The golden rules of any lifting operation are to know the load to be lifted and then to check the supporting structure is compatible. The failure of one rigging point on a lighting bar may cause a progressive failure – potentially very dangerous to anyone beneath during set-up, event or breakdown.

The venue has a duty to provide information and to an extent 'police' the use of their facility. The user (employer and employees) has a duty to work safely and to ensure that the loadings are within the design; it's no different to an electrician working out loads per phase. The difference is that in most electrical installations there are circuit breakers. Gravity doesn't use trip breakers.

SHOCK
KN KN KN KN KN KN!!

# Rigging Call #13

*March 2007*

### Secondary Suspensions

The issue of secondary suspensions for temporary rigging has long been a source of controversy. Often, the effort of rigging and de-rigging secondaries far outweighs the supposed level of safety they afford. The point of a secondary is to guard against a load falling; it must be said that very often they would not work because of the way they are rigged or the nature of the structure that they are rigged to. So why bother? Secondary suspensions are often a venue requirement, doubtless believed to be in everyone's best interest. I've seen riggers hang secondaries such that the lower end falls into the chain bag of the motor they were supposed to be 'safety-ing' to satisfy the venue that the "safeties were on". If you're going to rig them, you'll make sure they'll work, surely?

The origins of the safety are unclear: one theory suggests that house riggers were concerned about the integrity of the hoists or trusses they were asked to hang (sometimes with good reason), particularly when hung in front of the crowd barrier. An explanation I heard was that the audience weren't employees of the production and so weren't insured; the crew and artists were (and so could be hit by falling objects?). Another facet of the argument (from shows using pyrotechnics on trusses) was the fire safety aspect with regard to man-made fibre slings. Steel slings were seen as a necessary back-up for fibre slings on trusses and in the roof. Routing a wire rope sling next to the polyester one often meant there was slack in the steel and the force generated by the drop distance may have been enough to break the wire rope, and almost certainly create an excessive load on the supporting structure. Much better to use a steel sling, with softening as appropriate, and forget polyester slings for lighting trusses or applications where heat (maximum operating temperature usually quoted as 85° - 100° C) may affect man-made fibres.

So the question is, do we need secondaries at all? I would contend that we do, but only where a) a failure would otherwise cause a load to fall or swing (e.g. a truss hung on two motors, or a single point suspension or, b) the suspended load might be subject to other forces (surrounding activities such as use of plant or the collapse of an adjacent structure). In other words, do a risk assessment – pretty much what LOLER requires anyway. Any load must be supported adequately by equipment that is strong and stable enough, and that equipment must be "installed to reduce to as low as is reasonably practicable the risk of the equipment or load striking a person, or from the load drifting or falling freely, or being released unintentionally, and that it is otherwise safe". The phrase "otherwise safe" is explained in the guidance to Local Authorities thus: "It is not intended to be an absolute requirement and compliance will be achieved if the risks are reduced as low as is reasonably practicable".

Two British Standards relate to entertainment industry rigging (BS 7905 and BS 7906). BS7905-1 defines a secondary suspension as a suspension "capable of arresting the fall of an individual item of equipment when there has been a dislocation of its primary attachment to the lifting system". This obviously covers luminaires but could apply to lighting trusses. BS7905-1 also mentions "machines used to suspend loads over people" recommending "secondary suspensions should be provided before anyone passes below the load". This applies to hoists not designed for use overhead by the manufacturer. British Standard 7906-1 (Code of practice for installation, use and removal of above stage equipment (excluding trusses and towers)) specifies three distinct uses of secondaries:

1. Luminaire 'safety bonds' – necessary where luminaires or other equipment (e.g. loudspeakers) are hung on a single hook clamp or fitting which may be dislodged.
2. For fibre ropes or slings where combustion is foreseeable (e.g. a hemp set carrying lighting equipment) – and suggests that a steel wire rope secondary suspension is required.
3. Lifting equipment or anchorage failure ("assessing the likelihood of failure of any attachment point on the load and whether this would lead to a dangerous situation. If so, consider adding extra attachment point(s) to provide redundancy".)

The standard states that it is best practice to use multiple hoists, so that the failure of one hoist would not cause a dangerous situation. It seems that this concept of redundancy is the most sensible course of action wherever possible; it ensures a load is more effectively supported and avoids time spent at height trying to install taut suspensions, often against the clock. BS 7906-1 and the 'Yellow Book' (*Technical Standards for Places of*

*Entertainment*) agree that for permanent installations, "adequate safety factors are essential for all flying installations. Therefore, provided the equipment has been rigged and tested appropriately, secondary suspensions should not normally be necessary. If there are any doubts about the quality of the installation, it should be re-inspected by a competent person and rectified if necessary. However, the competent person may decide that a secondary suspension is required".

The point in the Yellow Book that concerns me slightly is the recommendation that secondaries should be used when equipment is rigged 'temporarily' (defined as 'when it will be decommissioned and re-used elsewhere'). I would suggest that where a temporary suspension is made using equipment subject to LOLER compliant inspection and rigged from positions identified as suitable by a competent person, then the use of secondary suspensions is not normally necessary, provided an adequate number of suspensions are used, particularly if the advised 8:1 factor of safety is used. The resulting span or cantilever and increased loads created if a suspension failed must be considered when deciding on the positions of suspensions.

So if a rigger decides that secondary suspensions are required, BS 7906-1 may help when selecting and installing them. These are:

- "Selected so as to be able to withstand any likely shock load. Secondary suspensions should have a
- SWL at least equal to the required SWL for the primary suspension."
- "Of limited combustibility, where required to provide fire resistance."
- "Securely attached." (Correctly rigged and at appropriate positions on load and structure.)
- "Arranged so as to minimise the fall of the load before restraint occurs. A maximum drop of 150mm is recommended."
- "Arranged so as to retain the load in the same orientation as established by the primary suspension so that the load will not topple over, swing or rotate should secondary suspension take effect."

Clearly, no hoist, truss or similar equipment should be used if thought to be less than fit for purpose, but where redundant or secondary suspensions could afford a reduction in risk and can be employed without disproportionate effort or hazard we should surely use them.

# Rigging Call #14

*May 2007*

## Lifting Plans

As we should all know by now, LOLER (The Lifting Operations and Lifting Equipment Regulations 1998) requires any lifting operation to be properly planned (Regulation 8(1)(a).

Most industries routinely use lifting plans; they are often a requirement of insurance cover, and certainly one way for an employer to ensure safe systems of work are used.

So what is a Lifting Plan? It sounds a bit industrial for us entertainment industry types. Usually, it will be a written description of a 'lift', detailing the features that may affect the preparation and method of lifting and lowering. The plan should address issues such as 'the lift' remaining safe for the whole of the lifting or lowering operation, i.e. from when the load leaves the ground to when it is landed again. The lifting plan should include details of the location and environment, the loads to be lifted, the machinery to be used, the method used to attach the load to the lifting machine(s) and so on. It may even go as far as to specify the level of training or skill required by the lifting team.

LOLER requires every lifting operation to be properly planned, adequately supervised and carried out safely. Using a lifting plan should ensure that the person planning the lift can communicate the necessary information to the person supervising the operation and in turn, that person can supervise those carrying out the work. Lifting operations are typically either routine or unique and range from the simplest of single hook straight lifts to much more complicated operations involving many lifting machines and complex load distribution.

Generic lifting plans for routine activities are a practical way of ensuring safe working methods are used, and a simple template can be used to make sure that all the relevant factors are considered. The plan becomes a method statement, a checklist and can be used for training purposes.

Routine lifts are lifting operations carried out on a regular basis, typically very similar events each time they are carried out and employ equipment specifically installed to perform that function. Using flying bars or winches permanently installed in a venue would often fall into this category if the equipment is designed for the purpose.

Complex lifts are lifts that, for example, require multiple hoists or suspensions or that are particularly heavy, awkward or are to be carried out in less than ideal circumstances. Complex lifts may become routine, but being complex still require a greater degree of planning and/or supervision. The use of a written plan enables off and on-site preparation, supervision to be effective and provides a schedule for the lift and lowering operations. "The aim is to ensure the lift is not cobbled together on site using potentially inappropriate resources."

There is obviously a strong connection between the lifting plan and a risk assessment; the method stated in the lifting plan will already have started the risk assessment process and the person preparing the risk assessment should certainly have sight of the lifting plan. The aim is to ensure the lift is not cobbled together on site using potentially inappropriate resources.

Lifting a touring truss system, a PA or heavy scenery requiring multiple hoists or winches may be routine on a tour or whilst in repertoire, but is nonetheless complex in terms of a lifting operation. Using a written plan during the production period may be essential, and even once it becomes an established procedure the plan can be used for reference and training new staff. Loading in and out of venues using hoists or other lifting equipment is also a routine task in as much as it may occur often, but may also be complex owing to the nature of the loads being lifted. This is more akin to mainstream industrial lifting and discipline is required to operate safely in the often less than ideal conditions encountered.

The person appointed by the employer to plan the lift (a 'competent person' is required by LOLER) may be an in-house employee or if that competence does not exist in-house, brought in as a subcontractor. That competence is derived from experience of similar work, theoretical knowledge of lifting operations and how lifting equipment is designed to be used.

Supervision should be proportional to the risks in the job and take account of the people involved: for example, an experienced flyman bringing bars in to hang cloths would not require supervision. Relatively inexperienced riggers preparing a mother-grid to be flown on perhaps 20 chain motors will require much more supervision, and the coordination required to carry out the lift safely requires one person in overall control of the operation.

Supervision does not necessarily mean the direct physical presence of a supervisor at all times: an experienced person, or a person sufficiently experienced to carry out the work in hand, would typically require a briefing or 'tool box talk' in order to understand the work planned.

To carry out the lift in a safe manner requires the combination of available resources, the plan and the appropriate level of skill to carry it out. Many entertainment industry lifting operations are highly specialised and experience of the industry and the way it works on site is important. Another important part of carrying the work out safely is checking the condition of the equipment to be used prior to use and when the assembly is complete prior to lifting.

The level of detail required in the plan should be in proportion to the task and the risks inherent in it. Significant risks require the plan to include detailed information so the work can be carried out in the right sequence and using the appropriate equipment as intended by the person planning the work. This will be based on many factors but should include the information featured in the adjoining table in as concise a way as possible. Drawings and sketches to illustrate techniques and the stages in the work are particularly useful; a great deal of information about position, loads and lifting method can be communicated on a simple drawing. The use of lifting plans should ensure work is carried out as designed. Without them, there is a real risk of improvising and whilst rigging is not an exact science, planning safe lifting operations is an essential task in the production process.

| EXAMPLE LIFTING PLAN (METHOD STATEMENT) | | |
|---|---|---|
| **Consider the items in this column** | **Specify in this column** | ✓ ✓ |
| Description of activity | | |
| Production/site | | |
| Identify competence required for task | *Lifting equipment, work at height, etc* | |
| *Person responsible for lift | *Name of person* | |
| Appropriate supervision and number of crew | | |
| Write/review risk assessment | | |
| **The load and lifting it** | | |
| Weight of load(s) to be lifted ascertained | *In kg, +/- 5%* | |
| Check inspection status of all equipment | *Current LOLER certification* | |
| Lifting points provided on the load suitable | *Constructor's authorisation* | |
| Supporting structure checked as safe to use | *Venue authorisation/structural report* | |
| Other work being carried out at the same time | *Activities of other contractors causing obstruction to the lift, etc.* | |
| Proximity hazards to people, equipment and structures | *Segregation from people and other work*<br>*Lifting people or loads over people – extra precautions* | |
| Lifting sequence to be followed from mark out to final trimming or landing | | |
| Signed off as complete (all rows ticked)<br>by Rigging Supervisor<br>by Production or other responsible person | | |

THUNDERBIRDS

# Rigging Call #15

*July 2007*

### Rescue Planning

What happens to someone who is suspended in a harness after a fall is not often considered? This is for several reasons. Firstly, the assumption that wearing a harness will somehow make the activity 'safe'; secondly, the way the equipment works is not understood (and it is often supplied without training); thirdly because the risks associated with harness suspension are not widely recognised and lastly because the likelihood of a fall occurring is perceived as remote.

There are several reasons why a fall may occur. An incidental loss of balance maybe caused by an injury or electric shock, slipping, or as the result of a more significant event, for example a structural failure during the course of construction or dismantling of a structure. Secondary injuries sustained during a fall also need to be considered; impact by structures or objects may be extra significant hazards. The Work at Height Regulations 2005 in the UK requires procedures to be in place whenever work at height is carried out. Employers must ensure anyone planning or carrying out work at height is competent to do so, and this includes considering what happens to someone who has fallen and is left suspended. Rescue is primarily the responsibility of the employer since it is the employer who has decided to use methods that could result in a fall. Rescue methods depend on the nature of the work, the workplace and the equipment being used. The emergency services may not be equipped to respond to people working in relatively inaccessible locations and it's not enough to assume the Fire Service will be able to deal with such incidents. However it is good practice to invite the emergency services to visit workplaces where work at height will be a regular activity so they are aware of site conditions should they be required to respond and so they can offer advice to employers.

So why the need for rescue? Falls whilst using fall arrest or work positioning equipment will result in the fallen worker hanging in a harness. Hanging in a fall arrest harness leads to unconsciousness in a relatively short time even if no injury is sustained, a condition now referred to as harness induced pathology or orthostatic intolerance.

Blood is retained in the legs as the result of 'venous pooling' caused by gravity, making it difficult for the heart to pump blood to where it should be. This leads to fainting and potentially to death if normal circulation is not restored within a short period. If the casualty is unconscious, rescue is required within about 10 minutes or the consequences may be serious. Casualties who are injured or unconscious should have their legs lifted such that their thighs are at least horizontal, assisting circulation whilst rescue is prepared and carried out. This can be achieved by using slings or rope passed under their knees and tied off to their lanyard attachment point. This simple act may be a life saver, particularly for unconscious or injured people. Conscious casualties should be encouraged to wiggle their toes to assist in circulation. Vigorous activity is unsustainable and may have a counterproductive effect, wiggling the toes can provide enough 'muscle pumping' to counter the effect of gravity.

Risk assessment highlights where fall arrest and work positioning techniques will be used and must also consider the situation from which the casualty may need to be recovered. The length and type of lanyard or fall arrest device may have significant bearing on the rescue methods that are appropriate for a given situation. In the entertainment industry, there are wide ranges of equipment and situations from which a casualty may need to be recovered, for example:

- ladders, fixed and temporary, with or without using PPE
- lighting bridges and catwalks
- lighting booms
- scenery and structures
- lighting trusses or grid systems
- lanyards (fall arrest or work positioning lanyards of webbing or rope)
- inertia reel devices (retractable steel wire or webbing)
- horizontal fall arrest lines (proprietary rope or webbing systems)
- vertical fall arrest rope systems (rope grab systems, such as the ASAP)
- horizontal fall arrest wire systems ('Latchway' type systems)
- fall arrest nets (often difficult to recover people from)

Each situation requires consideration which should include:

- anchors from which to raise or lower the casualty (e.g. not one that has just sustained a fall)
- the rescuer(s) safe access to the casualty
- the surrounding environment (e.g. dark, noisy)
- the condition of the casualty (e.g.nature and extent of injury)
- condition of PPE (e.g. can the harness be used in rescue or has it been damaged in the fall?)

If 'self rescue' (or an assisted self rescue) using slings or stirrups to get out of suspension and regain a place of safety is not an option because the casualty is incapacitated in some way, rescue by others is required. The technique entailing lowest risk is usually via powered access. If that is not feasible, non-powered access may be an option. Lastly (because it is typically more hazardous to carry out) 'rope rescue' may be needed.

There are three principal methods of 'rope rescue':

- attaching the casualty to a rescue system remotely, often from a position above
- rescuer descends to release casualty and continues in tandem with casualty
- rescuer(s) descends to casualty and rescue team lower (or raise) casualty and rescuer(s)

There are many considerations in rope rescue, the most important include:

- maintaining rescue equipment in good order and readiness
- how the rescuer(s) gain safe access to attach the casualty to the rescue system
- the first aid required before, during and after rescue
- using anchor points and equipment appropriate to the task in hand
- condition of the casualty's PPE before being used in the rescue
- being able to talk to and see the casualty during rescue
- edges causing severe abrasion and cutting of ropes during rescue operations
- avoiding 'shock loads' on all parts of the system and anchor points
- procedure on reaching the ground – how to maintain sitting posture and airway

- allowing rescuers regular opportunity for practicing rescue techniques

A rescue procedure might follow this pattern:

- The fall occurs, and the alarm is raised
- If rescue is required, immediately summon medical emergency services
- Decide which (pre-ordained) rescue method to employ
- Carry out the rescue in a controlled manner
- Monitor the casualty's condition (first aid provided if available and necessary)
- Advise the medical emergency services about the circumstances of suspension
- Ensure even an apparently unhurt casualty gets medical attention without delay
- Report and review the incident to improve methods and learn from the experience

Falls don't happen often, but when fall arrest equipment is used, a quick and effective rescue may necessary in order to prevent a potentially fatal situation arising. This may be a simple solution involving little equipment or skill, but the appropriate technique needs to be considered before the work starts and each time the method of work or the work environment changes.

> Originally there was advice printed here in good faith about lying casualties down once released from prolonged harness suspension. This advice was superseded in 2009 by the HSE's guidance note based on Dr Adil Anisesh's report which concluded:
>
> No changes should be made to the standard UK first aid guidance for the post rescue recovery of a semi-conscious or unconscious person in a horizontal position, even if the subject of prior harness suspension. No changes should be made to the standard UK first aid guidance of ABC management, even if the subject of prior harness suspension. An emergency 999 ambulance or equivalent qualified paramedical or medical provider should be called for anyone who becomes unconscious in harness suspension whether apparently recovered or not.

RIG OR DIE
07

# Rigging Call # 16

*September 2007*

## PLASA Show 2007

Here is a round-up of rigging related topics from this year's PLASA show at Earls Court. There were quite a few products that caught my eye.

The Vertex winch and control system from Vertigo Rigging is a natty wire rope winch system capable of up to 2m/second lifting speeds operated by simple joystick controls. Movement can be plotted in real time and learned once the cue or sequence has been established in a simple 'record' command. The move is memorised and can be played back using a very useful optional function to lose pauses in the original sequence to smooth out the effect. The winches are housed in a frame that allows mounting to grids, trusses or the ground and have lifting capacities up to 250kg with a 50m height of lift when using 5mm wire rope.

Two Revolution winches from Unusual Rigging were being used to fly a silver clad aerial artiste in regular elegant flying performances. The Revolution is a wire rope winch using a zero fleet design and extremely portable, being built into a general purpose truss module. It boats a unique 3600 lead off sheave meaning the wire rope can be run in any direction from the end of the winch – very neat indeed. The winch can lift 150kg with a 10:1 factor of safety for lifting people and for lifting objects (using a less demanding 5:1 factor of safety); the Revolution will lift 300kg at 3m/second over 30m height of lift using 6mm wire rope.

The 'Exe' hoist was being shown on the Litec stand. This is a development of the Lodestar chain hoist which includes some clever features such as a manual local override of the C1 hoist functions, allowing the hoist to run as a standard unit – useful when a hoist requires re-rigging, for example, but re-integrates with the system when the override is released. This is because each hoist has a unique address and each function of the hoist (as a basic hoist with D8 classification) is recorded in a Flash memory. The hoist runs at fixed speed (4m/minute) and has an LED display for error messages such as slack chain, data failure, etc. It uses bi-processor technology to provide constant monitoring of the hoist's performance with a redundant system as back up. The hoist monitors load, heat and current and gives users a clear picture of the overall health of the lifting system and control over hazardous situations as they may arise.

Beamhoist from Stage Technologies was on show, a product seen before but this time with its workings exposed. There's something very pleasing about its simplicity and robust design. Easy to install and with a very low profile it looks to me like a unit that will have many applications. Rigging centres can be as close as 150mm and it will accommodate up to seven lines. The bigger 600kg unit has a 0.5m/second with a 10m height of lift; the smaller 300kg model lifts at 1m/second with up to 20m height of lift. Interesting to compare this with the other two winches described above.

The use of fall arrest equipment when working on trusses is still of constant concern to me. Observing many exhibition build-up periods or the maintenance tasks carried out after a show has closed for the day will show any number of technicians using the right kit the wrong way. Many manufacturers are now warning users that their trusses are not designed to be used as fall arrest anchors. Exhibition stands constructed from trusses rarely allow the distance required for most fall arrest lanyards to deploy. Very often between 5 and 7m clearance is required, a stand no more than 4m tall is unlikely to do the job. It always surprises me that a technician is provided with, or has bought, around £300 worth of PPE but hasn't been trained to use it. The worst thing is they think they are 'safe'. Knowing the clear distance required into which to fall is a critical piece of information, as is the strength of the structure you are attached to. Even if the required clearance is available, there's every chance that in the event of a fall the truss structure would fall over.

The most significant advance in entertainment rigging doesn't involve equipment. It is the National Rigging Certificate. This is the first scheme in the world to measure competence in practical and theoretical entertainment rigging expertise. The NRC, administered by the PLASA organisation, was launched at the Show and candidates were being assessed in a specially constructed environment at the side of the exhibition floor in Earls Court 2. Visitors could watch the practical assessments, the administration and written elements were conducted away from the show floor by a team of qualified assessors and an internal verifier. The scheme is now up and running – contact PLASA to register. The first candidates to achieve their Level 2 and Level 3 awards were presented with their certificates in a ceremony on Monday the 11th September as were the first assessors to achieve the A1 award to assess candidates on the scheme. The assessors had also achieved the Level 3 award as part of their assessor status. The PLASA team, led by Nicky Greet, and the representatives of the rigging industry must be congratulated on making this happen.

At a more technical level, the WorkPro lifting towers caught my eye. The LW 475 R and LW 461 R are almost re-engineered Superlifts and seem to be well thought out, robust units with some nice touches like the protective guards at the upper end of each mast section. The LW 476 R will lift 220kg to 7.6m. There were also some similar, though smaller, units for lifting lighting that appear to be significantly better than many of the other heavy duty lighting stands on the market. To see a lightweight truss lifted at each end by little more than a wind up lighting stand has always seemed wrong. It is interesting to consider the controversy over double brakes on hoists, secondary suspensions for trusses and so on when the use of two flimsy ground supports goes unchallenged. There are probably more of these set-ups in use on a daily basis than any other method.

Rope Assemblies were giving away copies of their Little Black Book catalogue of rigging supplies and PPE. An interesting development from the company is their identification system for logging equipment – the constant battle to find a reliable method of electronically identifying metal items may be over. As a requirement of LOLER, lifting equipment and accessories need a unique means of identification. Rope Assemblies will offer an electronic label system which can be stuck to shackles and rings as well as slings or steels – sounds promising. In my experience using metal or plastic tags or barcode labels has never been particularly successful so it will be good to see this system on the market.

From a rigging point of view PLASA07 included some important new products. It will be interesting to report on progress after PLASA08.

NO THANKS
YESTERDAY
TODAY

# Rigging Call #17

*November 2007*

## Planning

"We have to do it this way because we haven't enough time to do it properly."

This statement is made quite often with reference to getting a show up on time, which of course is why we are all in this industry – the show, the event, the opening – people's entertainment. "The show must go on" – must it? "We've got to do it this way" – who says? Let's not forget that's what it is, an entertainment. It's a show we're working on; not a heart transplant. So why is it that we have such a deep seated desire to get the show on at all costs? Very often we are trying, diligently, to achieve the look the designer or director requires – but at what cost? Getting the gobo sharply focused for a scene in Act 2 actually doesn't equate with someone scampering unprotected along a truss during the interval. It is not uncommon to be asked to work to a schedule that means in the time allowed work will have to be carried out in a way that does comply with legislation. Ironically, often the hours worked don't comply with legislation. Too few hours to do the work safely, and too many to make working safe!

So work carries on. A combination of loyalty to the rest of the team and the audience, a love of the business, a CV to nurture and not wanting to seem less than a hardened professional? The only people who know what the show *should* look like are us, the people presenting it. How often has a friend whose show you have just watched, been really embarrassed about some technical gaffe – one you hadn't even noticed?

That's not to suggest for a minute that we shouldn't try hard to produce the goods, but neither should we feel that life and limb should be risked in the process. It's easy to decide how the job should have been done after the ambulance doors have been shut.

It is often an individual's reputation and the desire to be somewhere else that has driven someone to get a job done at all costs. Very often the work can be done more quickly in the 'traditional' manner and the individual can indeed get the work done, collect the pay cheque (so to speak) and get home early. There have also been some situations where an individual cut corners and didn't get home at all, with or without pay cheque.

There have been several work at height fatalities and serious injuries in the last decade as most of us know, and to some this was a wake up call. Those who have been working at height for some time typically fall into one of two categories, careful and lucky. Most of us have witnessed lucky escapes, sadly for some the not so lucky ones. But the vast majority of people in the industry have not been touched directly by it and it doesn't serve as a reminder to them.

We are working in a job we (presumably) love and because of that many of us are often prepared to accept greater risks than in other industries. Cast your mind back to the early days in your own career as a technician and tell me you didn't climb on things you wouldn't now. Very often ours is a youthful industry, the work at the coalface being carried out by young and enthusiastic people who are keen to impress and earn a reputation for getting the job done with a view to a lucrative career.

This means work is often carried out in a manner that may rely more on agility than forethought. It's also true to say many of those in their 40s and 50s who are now planning and supervising work were rarely aware of any regulatory constraints until comparatively recently. Consequently, they may not be supervising the work in line with current legal requirements. In pre LOLER or Work at height Regulations days there were regulations that applied to entertainment rigging but not many in the industry were aware of them.

The Health and Safety at Work Act came into force in 1974 – and its requirements were the same then as now. However, the way work was planned and carried out could have in the event of an accident left employers open to prosecution in many cases. In some ways not much has changed, except that since the 70s I suspect the number of live shows and events being put on has increased significantly. Also, think of the many hundreds if not thousands of recorded events that take place every year with all the lighting, sound and scenery that it entails, including rigging. That means more technicians are employed and therefore a greater risk of accidents exists simply because more people are, for example, connecting mains, working at height or operating plant and vehicles.

It would be interesting to know if the number of accidents has increased in line with the increase in shows. I suspect it has not; as we often say, the industry is actually pretty good at looking after itself. When a fatality does occur, it therefore stands out far more prominently.

I would contend that for most technicians, the hours worked are far more likely to cause accidents than virtually any other factor. The Tallescope issue is another case in point. To continually ascend and descend a Tallescope ladder is cripplingly tiring; having enough suitably skilled technicians to be able to rotate focusing duty is highly unlikely in most theatres. The hours spent driving to and from shows as part of the working day is another issue. There are Regulations about both of these matters, but they are routinely ignored by many employers and self-employed people.

There are a large number of well-planned and organised shows and events and the number is growing, but there are still far too many that are not.

So the technician is often in the position where the demands of the show, pressure from the boss and not wanting to let the audience down weigh heavily.

Of course the show should go on, but it may have to look and sound less than ideal because the budget won't run to extra crew member, hotel room or access platform.

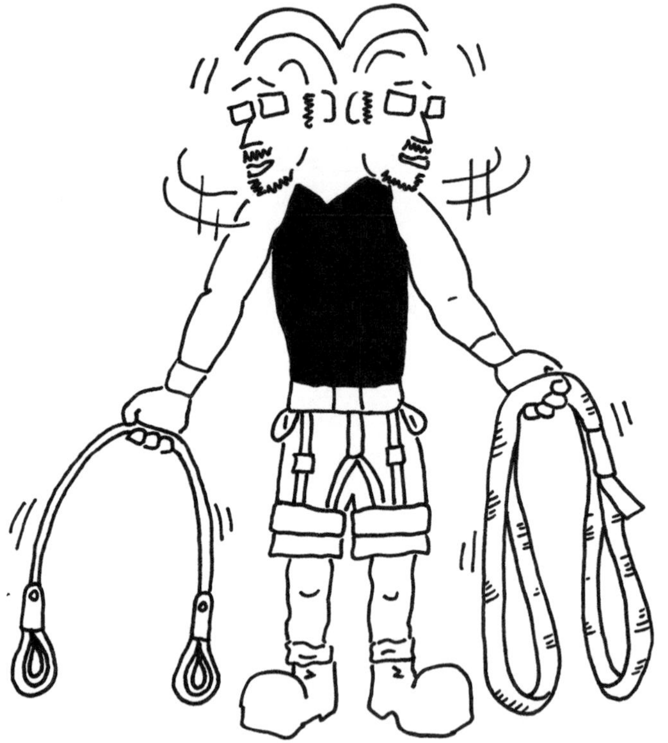

# Rigging Call #18

*January 2008*

### Roundslings

This time, in our bi-monthly rummage through Uncle Rex's rigging box, we turn our attention to our old friend the roundsling, or 'Spanset' as it is often known. Spanset Ltd must be very grateful to the millions of users of polyester roundslings who refer to any roundsling as a 'Spanset'. Oops, did it again. Roundslings are found in every sector of industry involved with lifting, from machinery bumping to aerial acts. The principle is a very old one; staple fibres of old ropes were recycled and laid up into endless slings and 'marled' with twine to shape and protect them; the modern day version uses polyester yarn and a woven sleeve of single or double thickness webbing.

Roundslings are available in a wide range of lifting capacities and lengths (these days referred to as Effective Working Length (EWL) – rather than the traditional circumference. The Working Load Limit (WLL) is denoted by sleeve colour and markings, confirmation being obtained from the identification label attached to the sling. The entertainment industry typically uses roundslings with safe working loads of 1000, 2000 and 3000kg, denoted by violet, green and yellow sleeve colour respectively. In addition, the sleeves often have a line stitched or dyed around the circumference to denote the WLL; one line for 1000kg, two for 2000kg and so on.

Manufacturers often embroider the WLL or company name into the sleeve providing a highly visible and hardwearing means of identification. Like all CE marked lifting equipment, roundslings should be supplied with instructions for safe use – read them. This article is not a substitute for that and only looks at a few key points. So, to the pros (and cons). They are remarkably strong and in relation to their strength very lightweight. They have a factor of safety against failure of 7 against the minimum failure load and should only stretch around 3% at their safe working load. Roundslings are soft to handle and very kind to the load or as is often the case in our sector, the supporting structure.

Paint finish on beams in exhibition halls won't be damaged easily by a roundsling. Because of their construction, roundslings spread load over a wider area than a steel sling making them ideal for slinging trusses or other structures with slender proportions. In a similar way, wear to the sling is spread over its full circumference because it has no 'eyes'. Fixing a roundsling in position means it will always receive wear in the same places and thus should be avoided. From a rigging safety point of view, if a roundsling gets dropped it will cause less injury than other slings and is itself unlikely to be damaged on impact.

The 'cons' are equally important and there are some real no-no's associated with using roundslings. The most likely is knotting or choking on too small a diameter. A knot does several things to a roundsling but the most important is that the fibres are severely bent over a smaller and smaller radius as the knot tightens, weakening them and preventing them from being equally loaded. This also action also generates friction thus heat which can weaken the fibres. So don't 'lark's head' them together – use a shackle.

Using a choke hitch *around a load* is permissible, but for the same reasons, the user should de-rate the sling by 20% in these situations. Keep the choke angle to less than 1200 at the choking point. When using shackles, hooks or rings with roundslings, aim to use a fitting wide enough to let the width of the sling lie more or less flat and over as large a diameter as possible. This ensures the fibres are running parallel and over a kind edge to allow the fibres to work efficiently. Pinching as a result of shackles used in 'open basket' configurations, or being forced into small shackles will have a similar effect. Wrapping 50mm truss chords will not adversely affect the sling's strength, that would start at less than half that diameter but care should be taken to avoid roundslings tightening on themselves when lifting heavy loads. Similarly, avoid twisting roundsling fibres as they pass over edges to optimise fibre strength.

Other things to consider include poor resistance to cutting on sharp edges – often it is worth using softening or packing to protect slings, particularly when the sling could be pulled at an angle along an edge or into an angle. In response to demand from the entertainment industry some manufacturers supply roundslings with a black outer protective sleeve, which can lead to confusion over SWL. Users should always check the label before using the sling – there have been slings supplied with black sleeves with two lines around the sling which were actually 1000kg slings though in most respects they were indistinguishable from 2000kg slings.

Good practice requires a pre-use check not only of condition but also of load capacity. Unfortunately, many users expect a black roundsling to be a 2000kg sling and a close look is needed to distinguish that it is so. Manufacturers have to assume you will lift up to the SWL with their products. In the details above, the same assumption is made. However, very often in our sector the lift is nowhere near the capacity of the sling, it is being used for convenience and will be supporting only a fraction of its SWL. This is not to say that it doesn't matter – be aware. However, when a 1m roundsling is choked on a counterweight barrel to hang an internally wired weighing 50kg, many of the considerations detailed above are virtually irrelevant. At the other end of the scale, a load approaching the sling's SWL requires careful consideration and sling placement and connections must be well made.

Local heat from luminaires, sparks from pyrotechnics, welding, grinding or similar activity could well be a significant hazard when using roundslings. As far as heat goes (as distinct from sparks – burning particles), most manufacturers state 100º Celsius as being the maximum operating temperature for their slings. This may vary from one manufacturer to another, so check when ordering. If risk of being melted is foreseeable, use an alternative sling material – chain slings if you really want to be 'fireproof'. There are roundslings with aramid fibres that are designed specifically for use when welding and also the wire rope cored slings such as the Softsteel and GACflex slings.

Incidentally, it's worth repeating information from a previous issue – roundslings are rarely affected by UV in the same way as a flat webbing sling might be. The loadbearing fibres in a polyester roundsling are, according to one large manufacturer, protected by the sheath and they knew of no instance where UV degradation had taken place. And finally, never use a roundsling if you can see the white load-bearing fibres through a hole in the sleeve. So obvious I missed it from the first draft of this article!

# Rigging Call # 19

*March 2008*

## Tallescopes

The tone of this column is often 'safety' related and no apologies for that; it is important to recognise that working in this industry can be potentially hazardous. Potentially. Three phase electricity, ladders, machinery, falling objects, pyrotechnics, work at height, replica weapons – the list goes on. There is an issue that affects a great many technicians in the theatre and wider entertainment industry that just won't go away and that is about moving Tallescopes with someone in the 'basket'. The only reason that this is done (no-one enjoys it) is because it reduces the need to climb the ladder to a large degree, reducing the likelihood of a fall and speeding up the work and reducing exposure to the hazard at the same time.

That's why there aren't many accidents with them… A few years ago, a local authority officer was inspecting a venue and saw a Tallescope being moved whilst occupied. This 'mobile' use goes against the manufacturer's original instructions. To cut a long story short, when they were consulted the HSE decided that 'mobile' use is a breach of PUWER (The Provision and Use of Work Equipment Regulations 1998). This was communicated to the Local Authorities. This column is not seeking to encourage people to breach Regulations, far from it, however following the manufacturer's advice when using a Tallescope in our sector would appear to increase the likelihood of a fall.

It should be remembered that there may be several hundred luminaires that require focusing in one hopefully short session. For 'mobile' use the Tallescope must be set up correctly (including the outriggers being fitted at right angles to the long axis with their feet just clear of the stage surface) and the operation carried out in accordance with a method statement detailing planning the work, staffing levels and control measures to be taken. It is absolutely understood that the need to focus lighting does not equate with the consequences of a fall from height. However the work needs to be done and also to remain commercially viable. Injuries sustained in falling from a ladder don't equate with the need to paint a window frame. However it is accepted that light duty work can be carried out from a ladder for a short period subject to correct use of an appropriate ladder.

It is a matter of fact that in many theatres the Tallescope is the only practicable safe method of carrying out a number of tasks at height. This is not to say we should not consider alternative means of access or cling to the past, but for the present in many venues the Tallescope is often the best tool for the job. The nub of the issue is that the manufacturer states a Tallescope must not be moved with someone in the basket. The manufacturer of an access machine designed for maintenance tasks in a wide variety of indoor and outdoor situations *would* say that. The Tallescope wasn't designed for use the entertainment industry. Neither was it designed for 'mobile' use (someone being moved in the basket) but many, many years of trouble free daily use in the vast majority of British theatres has proven it to be a very useful tool for focusing and many other tasks. There have been some accidents, but these all seem to have been caused by incorrect set-up or maintenance, and not by mobile use.

Mobile use of Tallescopes (to remain in the basket and be moved carefully in a planned manner from luminaire to luminaire) is far safer than continually going up and down a vertical ladder 100 times in one focusing session. It would a foolish technician that stayed in the basket whilst moving long distances, but no-one is advocating that they do. It would be a foolish supervisor that allowed the crew to use a Tallescope if there were safer options available. It seems to me this situation has arisen simply because the manufacturer's catch-all instructions have been assumed to apply to all situations. If the mobile use of Tallescopes is to be 'banned', then perhaps free-standing ladders (another universally used means of access in theatres) should also be banned? They are a lot less stable. That's not to say use of powered access should not be considered, and a great many theatres already make use of the vertical, single person 'IWP' style of machine where they can. However there are a large number of theatres with no reasonably practicable method of getting powered access onto stage. Even if they could the stage is unlikely to be strong enough for a scissor-lift or boom type machine. The largest Tallescope weighs 130kg.

I think the HSE will eventually come to the same conclusions as many theatres already have and review their position, but what of the future?

We can't rely on the Tallescope for ever. We should use the next few years to find alternative methods of work for that traditionally carried out from Tallescopes. This should involve set and lighting designers collaborating to make lighting automated where possible, bars that can be brought 'in' for maintenance – ideally avoiding the need to work at height at all. Providing means of access, platforms and bridges as part of the set infrastructure wherever possible is another option. Powered access would be the next best choice. Failing that, the Tallescope would still often be the best choice assuming a mobile alloy scaffold tower presents too big a footprint. (It should be borne in mind that moving an 'occupied' tower is also contrary to the manufacturer's instructions).

The extension ladder will probably remain the most versatile means of access in a theatre. If set designers are contractually obliged to consider safe work at height it could improve matters significantly. Too often there is an assumption that it is the 'venue's problem'. There will probably never be a single solution but the industry should be able to reduce the need to use Tallescopes. However, when it is the right tool for the job, we should be able to use one. The hours of work expected of technicians would seem a far more legitimate and pertinent target for authorities seeking to reduce risks.

# Rigging Call # 20

*May 2008*

### Loads on tents

Planning, or the lack of it, is something that we all moan about. We aren't given enough time, the budget won't stretch to that, Plymouth back to back with Aberdeen? The list is endless. On the other side of the coin, on the rare occasions we are given time, we rush the job in an attempt to get away early. The missing link is often at the start of the process, the promoters and producers need to involve technical people at the very start as part of the feasibility of a project.

Now more than ever, suppliers and technicians are better informed and can offer sound advice to organisers or producers that may save them money and/or time. The perception seems to be that if you ask the 'technical people' to get involved, they price the job out of the water. Not always – they want the work, but they also want to be able to do it properly and without working 24 hours a day. It may sound as if I'm 'Production Manager bashing', I'm not; the technical folk need PMs as much as PMs need technicians.

However, producers and promoters often seem to assume that because it was done in a particular way when they were on the factory floor (so to speak) it still is. Moreover, the perception that the requirements of the health and safety police or new-fangled methods will cost more may also be incorrect. The Production Manager also may not be au fait with technical progress – it worked that way last time, we're doing it again this time.

At this time of year, working in tents springs to mind. They come in many shapes and sizes from a simple two pole traditional marquee to the 'Kayam' style structures which can cover entire counties. Tents and tensile structures are often used when an event takes advantage of a site for aesthetic or practical purposes – they are generally quick to build and remove and have a certain novelty value. So the organiser has booked a site, contracted a tent supplier, put in seats, carpet and catering and is now thinking about the show or event itself.

Ah. No stage/lights/sound/blackout/power. This is going to be expensive. Large scale events generally use bigger suppliers and have bigger budgets and someone will come up with a cost effective solution that ticks all the boxes, generally a trade-off between production quality, value for money and wow factor. Call it planning. Smaller productions often use a one stop shop who may have little technical expertise in 'found' venues. The production company will typically use a technical supplier of equivalent experience. Now I'm not being rude, but a small lighting hire company is often unaware of what they are taking on rigging-wise; it's not what they do.

The typical loading for a 30m aluminium 'clear-span' tent seems to be roughly 200kg per portal frame. If a 20m lighting truss including all the rigging, the truss itself plus the payload weighs 200kg, uniformly loaded and picked up at 5m centres that will load the rigging points at approximately 20kg at one end, 56kg, 46kg, 56kg and 20kg again at the other end respectively. That's maybe five 'movers' or ten profiles, assuming 25kg for the truss and it being rigged direct to the tent. That's why people overload tents – there isn't enough capacity for what is perceived as a 'decent' lighting rig for a 12m wide stage. What should happen is that the lighting designer/rental company speaks to the organisers to determine the allowable load of the structure before the tent is booked. Strengthening may be possible, extra portals, frame bracing and use of different connections can be used. There are still limits of course – it still may not be 'enough'. There's a lot to be aware of working in tents apart from sheer strength – stability, wind loads and so on.

The fixings used to attach rigging to the structure are equally important. The best place to get advice is the supplier of the tent. I know a lot of people are reading this and saying "well the supplier was no help at all, they didn't know anything". Well if you ask the guy putting the tent up, he probably won't and even if he did, asking on the day you're rigging is leaving it a bit late and is unfair to all concerned. There a number of companies who rent 'kader clamps' rated at as much 150kg SWL to fit the extruded aluminium tent rafters. This is significantly more than the loading figures quoted for the rafters by the tent manufacturers. When challenged, I have had people tell me that the allowable load they can hang is 150kg because that's what it says on the clamp. Think again. Other common rigging methods used in tents include wrapping a roundsling over what are perceived as the 'purlin' ends. There is a problem with this. Very often they aren't purlins at all, but wind braces – stiffening for the portals. Look at the riveted clips used to hold them to the rafters.

Do they look like reliable connections? Ask the supplier where the best attachment points are.

In preparing this article I called a few tent suppliers and they all said they could supply a high tensile bolt to make the attachment to a rafter and that the allowable load was likely to be in the region of 50kg. (Do not take this figure as the answer – it will vary enormously depending on many factors, not least the tent span and prevailing weather conditions.) The allowable loads may be perceived as very small, but that is for a reason, particularly when taking into account the size of tent, the location and aspect of the site, and the time of year. A sure fire way of establishing if a tent is being loaded uniformly is to look at the ridge from outside.

My friend, colleague and structural engineer in New Zealand has a wealth of experience with the stability of these aluminium buildings on film locations where severe weather is likely. He recommends sighting the ridge as level across the whole structure, meaning an even loading has been achieved across the portal frames. This view was supported by all the suppliers I spoke to – a significant dip caused by a local overload may mean sudden failure in even a small gust of wind. An independent ground supported structure inside the tent is another option – but don't forget to check the load bearing capacity of the floor or ground that the towers sit on and to leave a gap between tent and structure to allow for tent movement. Every tent supplier I spoke to in preparing this article said that they were two or three steps away from the end user. Their customer is often an agency or event organiser who hires a production company who hires a lighting company who design a lighting rig to install. One manufacturer told me he 'wished these companies contacted them sooner because there are often easy additions they can provide to increase the load capacity of their structures'. I was told that often when the lighting has been designed and found to be far too heavy, the production company (the structure having been ordered by the event organiser) contact the tent supplier to carry out strengthening work.

At this stage in the process, the increased cost in strengthening the tent is often not considered viable. Unfortunately this often doesn't seem to prevent the lighting from being hung anyway. There are always loadings available, yet it seems that lack of forward planning and perceived 'unnecessary' expenses are the reasons that tent suppliers aren't consulted. Sadly none of this will get done 'on the day' – which brings us back to planning again.

# Rigging Call # 21

*July 2008*

### Terminations

Terminating wire ropes is something many technicians regularly need to do, and it's common to see terminations made incorrectly, often at the pick-up points on flown bars. Although we rarely hear about bars crashing to the stage, terminating wire ropes correctly aids safety for all of us, particularly in permanent installations. The most 'modern' standard is European Standard EN 13411 'Terminations for Steel Wire Ropes – Safety', published in 2002. It is in seven parts at present, covering thimbles, splices for slings, ferrules and their installation, socketing with resin/metal, wire rope grips and both asymmetric and symmetric wedge sockets. It is helpful in that there is one Standard that covers all the termination methods in general use. This is a Standard, not a Regulation; though best practice would call for the use of the most appropriate Standard.

### Wire rope grips ('Dogs')

Wire rope grips can be bought in several flavours depending on the efficiency required for the application, the rope type and its diameter. Using fittings in accordance with EN13411-5 provides a termination efficiency of 80% of the minimum breaking load of the rope. There are specific details on the number of grips required, the spacing between them and the torque required to tighten them.

Readers familiar with the DIN1142 standard will be glad to know it's very similar, so you won't have to re-learn much. The most common mistakes with dogs are using the wrong type or number, over-tightening them, and fitting them wrongly on the rope. The number of UK venues with four DIN grips on a 6mm rope is surprising (you only use three according to both the old DIN standard and the newer EN, though the new standard only provides a termination efficiency of 80%, so 5% less than the DIN requirement).

Many installers don't use torque wrenches on every dog, but it's worth knowing what the required torque feels like – it's less than most people would assume. Getting the termination right includes making sure the rope is straight, so the dead side isn't taking load off the live side. Load in the rope will always take the shortest path, so it is important to ensure it goes down the live side. The golden rule is check the manufacturer's instructions.

## Wedge sockets

There are two main types of wedge socket, each with some derivatives. Essentially, these are the symmetrical and asymmetrical sockets and the difference is as the name suggests. The two types are used differently and shouldn't be confused. The symmetrical type is the most widely seen in entertainment industry use; the asymmetric type is typically used on cranes and winches. It's common to see the correct method of fitting for a *symmetrical* wedge socket being used on an *asymmetric* type. Manufacturers and Standards are clear about how to do it: clamp the ropes together on a symmetrical one and not on an asymmetrical one.

Grips should be used according to the usual convention – the live rope goes under the saddle, the U bolt goes over the dead side. Get this wrong and the grip will act as a stress-raiser and quickly damage the wires. The reason the grip is there at all is to stop the wedge being displaced when the load comes off the rope. Symmetrical sockets are better for wires that are always loaded – for example passenger lifts (supposedly where the method came from), where the car is always on the rope, or in our world on flying bars and suspended loads such as acoustic panels or scenic pieces. Even though loads may increase (hanging something on the bar) there is always a load present (the bar itself). Asymmetric sockets are better for ropes where the load is completely removed – so for us point hoists, winches and so on. The rope will change position inside the socket and therefore the asymmetric type is designed to wedge into the socket as the load is applied. Some types have a grip built-in, eliminating the need to use a loose grip and doubt over which way round to fit the rope. The best advice: use the manufacturer's instructions and always identify the type of socket being used.

## Strandvise 'Bullets' or 'QuickTerms'

Bullets were originally guy terminations or repair splices for wire ropes used in the overhead power line business. They are simply a set of jaws that grip the wire rope when the rope is pulled, working much like the chuck on a drill. They are popular because they are quick and generally reliable. They are available with a CE mark, meaning they have been tested against the requirements of the Machinery Directive and may thus be used for lifting. When tested, the rope used was of fibre core construction and this is the best rope to use them with, certainly if approaching the SWL. The tests included ropes that were greased and it may be a surprise that no 'back up' dog was used. Most people lead the dead side up to the standing part before the Bullet and 'dog' it back, forming a loose loop. As with wedge sockets, this is not always appropriate: the rope grip could act as a stress-raiser on the loaded rope.

Bullets are made largely of aluminium and are therefore easily damaged. The 'bail' – the steel suspension 'hoop' – should not be bent or twisted, and the body should be free of dents that could affect the free working of the spring-loaded jaws. A common complaint is that they 'lock' and cannot be easily adjusted. That's because they weren't really designed to be! The way to release the jaws is to unload the rope, and with a soft hammer strike the base of the body in line with the direction of pull. This will release the jaws' grip; then insert a thin screwdriver or spike to compress the spring, slacking the jaws completely and allowing the rope to be moved. It does mean taking the load off the rope, which is not always simple or safe. If you try to lever the jaws without slacking them, you'll damage the aluminium body and bend your spike …

There are traditional workshop methods such as mechanical splicing (pressing alloy ferrules onto the rope) and 'socketing' where molten metal (these days resin), is poured into the lay of the rope, making a captive plug inside a metal terminal fitting. Newer methods include sophisticated compression fittings and various automatic devices used in suspended ceiling grids. Some offer TÜV certification and CE mark, some do not. The maxim should always be to choose the method most suitable for the situation and with regard to the risk to people and property in the event of failure. As ever, if you're unsure – don't do it.

NRC - LEVEL 2
QUALIFICATION
NRC EXAM
ACHIEVEMENTS
FORCES
KNOTS
NO
SWL
LOADS
NO

# Rigging Call #22

*October 2008*

### The National Rigging Certificate - some myths dispelled

With any new industry qualification rumours start to circulate – this time Rigging Call discusses what the National Rigging Certificate actually entails. The National Rigging Certificate (NRC) is the recognised qualification to attain when working in the field of entertainment rigging. It is not just about certification. The NRC is about the recognition of skills that are vital to the rigging industry, which are often overlooked. In the future the NRC will provide a route for people wanting to get into rigging as a trade. Previously this has not been possible since rigging has never been properly defined.

**The NRC is already raising the profile and awareness of the rigging profession and acknowledging the skills required**. The qualification is awarded by PLASA Qualifications, now an awarding body in the same way that City & Guilds or Edexcel are. PLASA are therefore regulated by Government in the architecture and delivery of the scheme and quality assurance is paramount in the system. The cost for each qualification (Level 2 or Level 3) is in two parts, registration to PLASA and then to the Centre, to cover the assessment itself. This is a one-off payment and includes one re-assessment if required. **There is no annual cost to the NRC**. On achievement, the candidate receives a certificate for the qualification and can then apply for the NRC ID Skills Card. The qualification is for life, but in order to allow the industry to self-regulate and ensure riggers are meeting current practices the Card expires after five years. This renewal will only involve assessment if there are any significant changes to legislation, techniques or equipment during that time. It does not mean the whole assessment process has to be gone through again. There will be a fee attached to the re-issue of the card and any assessment that may be required.

Assessment takes into account knowledge, skill and experience in equal measure – so simply being good at climbing or physics is not enough – it is the combination of all the essential skills that make up competence. **The assessment is based not on opinion, but on National Occupational Standards** (statements set by the industry of what you must do and what you must know to carry out a particular work activity). These act as the benchmark by which the assessors make their decisions. Assessors are monitored by qualified Internal Verifiers, based at the assessment centres, which are also verified by the awarding body making it a three-tier verification process for the candidate. There is a detailed appeals and grievance procedure to protect candidates in the unlikely event of dispute. At present the qualifications available for 'existing' riggers are at two levels.

Level 2 is for riggers working independently but not fully supervised, Level 3 is the qualification for someone who supervises riggers. **These Levels represent the minimum core competencies required for rigging and are just as applicable to the theatre, concert touring, event, film and TV, and anywhere entertainment or event rigging is carried out**. At Level 2, assessment covers the methods generally used in entertainment rigging such as marking out, preparing rigging assemblies, using PPE, safe systems of work to install rigging points, using motors, assembly and slinging of aluminium trusses, lifting safely, rigging secondary suspensions and making steel wire rope terminations, de-rigging safely and pre-use checks of lifting equipment and of PPE.

At Level 3, Centre assessment includes the practical assessment as per Level 2 candidates, and in addition includes carrying out a risk assessment and sitting a longer written paper. Following achievement at the centre, candidates also undergo an on-site assessment at a real job that must include activities such as supervising riggers working at height.

**After 2010, this 'Existing Rigger Route' will no longer be available and candidates will have to go assessment at Level 2 before going on to Level 3**. The new entrant route is being discussed by the Rigging Advisory Group and may involve formal training or an apprenticeship, as well as work experience, before the final Level 2 assessment. The direct access route allows experienced riggers the opportunity to be assessed at the appropriate level without working through the levels, but this is dependent on their current role. After 2010, all individuals will have to progress through Level 2 assessment to then undertake a Level 3 assessment. The body of knowledge required is contained within the comprehensive NRC handbook and candidates are sent this when they register with PLASA Qualifications. At this point they are also given a candidate reference number that stays with them and is eventually put on their NRC ID card.

So what will happen to you, the rigger, as a candidate? When you turn up at the Centre for assessment, you will meet your assessor, an experienced rigger who holds NRC Level 3 and the National A1 Assessor qualifications. The assessor will also have had specific training in the NRC scheme. A one-to-one session will ensure you are satisfied as to how the assessment will be carried out. You then sit a written question paper for an hour or so, which will be marked before any practical assessment. This allows the Assessor to establish whether there are any knowledge gaps, and so further question the candidate during the practical assessments, as some individuals find it easier to provide answers when applied in a different context. At all times the assessor is trying to establish if the candidate meets the underpinning knowledge in the Occupational Standards. After a lunch break, you'll have a practical assessment with the other candidates. The number in your group won't exceed four, so there is ample opportunity to demonstrate your own ability as well as showing you can work as part of a team. The scenario will be a rigging job that involves most of the tasks you'd expect; rigging motors 'up' and 'down', working off catwalk on a beam, slinging, using trusses correctly and managing chain motors and controllers.

There is a feedback session with the assessor before breaking for the evening to enable candidates to know what they have achieved so far. On day two there is a second practical assessment, creating opportunities to demonstrate that you can not only read drawings and mark out from them, rig points and lift safely, but also to terminate wire ropes and use safe systems of work including using PPE where this may be necessary. At the end of the assessment there is a review of the two days and you will be advised of your achievements so far and what remains to be done.

**There is no pass or fail – you achieve or do not achieve the required standard on different areas within the assessment. A candidate only requires re-assessment on the areas they did not achieve on** and have the opportunity to discuss the next actions with the assessor and will be sent an action plan to work from. Don't forget the scheme allows up to two years in which to complete from when the Centre assessment took place. For existing route candidates, it is possible to achieve Level 2 following the two day Centre assessment – providing all your evidence meets the standards. At Level 3 it is possible to achieve after the two day Centre assessment plus an assessment of supervision at a real rigging job – again providing all your evidence meets the standards.

The best description of the NRC? It's about what you do, not what you don't do. If you're a jobbing rigger, the assessment is about what you do every day. More information is available at www.plasa.org/pd/qualifications

STOP

# Rigging Call #23

*January 2009*

### Trusses

Trusses are used all over the world in applications ranging from concert lighting systems to museum displays. When used correctly trusses are a useful tool offering advantages over more traditional suspension systems, but when used incorrectly they can add risk to our work. Here are some common points of discussion.

### Distributed loads?

The loads on the supports of trusses supported on two points can easily be determined by calculation. Load distribution on trusses supported in three or more places is not so easily determined. Just because the truss looks level it is highly unlikely to be sharing the load between the available supports even where the truss is uniformly loaded. The belief that a uniformly loaded truss flown on three hoists will result in each hoist lifting 33% of the load is unlikely to be the case. The theoretical distribution can be calculated or even taken from tables, but in practice it is rare to find such a calculation is accurate. Even with the truss supported at the centre of three equal 'panels' the load will not be uniform at each support. The load at each support decreases as the number of supports increases, but the load will still not be approaching a uniform distribution across the supporting structure until perhaps six or more supports are used.

Circular trusses are a common lighting design configuration and require careful analysis. A truss curving between support points is a cantilever and could be creating large forces in the truss. Using plenty of supports reduces the span or cantilever, making the truss stronger and safer, but still this may not distribute weight equally. Truss grids carry more weight towards their centre when the grid is uniformly loaded for the same reason. A grid flown on nine motors (as three rows of three hoists, for example) will result in the centre hoist carrying in the region of 25% of the total load, those around the perimeter less than 10% each. Loads applied to a grid are rarely uniform and the loading at each point, thus on the supporting structure, is even more difficult to calculate. Load cells are definitely worth considering in all but the most simple of systems.

## Assembly – which way up?

Trusses are essentially a bundle of aluminium sticks welded together to make a structure than can carry weight over a distance safely and without undue sagging.

Most trusses will have a 'right' way up and the manufacturer's information should tell you which that is. This relates to triangular trusses and truss circles as much as to simple spans. A hiccup in the regular zig, zag, zig bracing pattern is often referred to as mis-noding and thought to be weakening. Mis-noding is actually the distance the forces in the member have to travel between the points at which members meet rather than the braces not running in sequence. Picking up a truss at a node point is usually desirable, possibly essential. One reason relates to mis-noding – applying a force (the pick up) away from the theoretical node point is itself mis-noding the structure.

Triangles are structurally efficient shapes, they don't move easily. If you can join lots of triangles together, predicting the forces in each member and sizing each accordingly you can make an efficient structure. If a triangle changes into a four-sided structure because the lines of the triangle don't meet at a point (so called mis-noding) a 'weaker' spot is created unless the design takes account of that. If the triangles are formed by a vertical 'where a diagonal should be', it is still a triangle and although it may not look so good the strength may be unaffected. The manufacturer will provide information about this in the user manual.

## Fit to use?

The truss and its connectors must be in good condition. Any loss of material must affect the load carrying capacity of the structure. Any weakness or inefficient transfer of load acts as a stress raiser and is potentially a point of failure. Local deformation or bending in a member could have a dramatic effect on the stiffness of the truss which in turn can affect the properties of other members. Lose a diagonal brace and the truss could collapse under a combination of circumstances. A badly bent or missing horizontal brace may allow the truss to twist or sway and produce a similar effect.

Always make a visual check of trusses before use and if it doesn't look right, it probably isn't – don't risk using it.

## Connections – weak spots?

A truss needs to be transportable so the bundle of aluminium sticks is made in small sections. Each section needs to be joined to the next with no loss of strength so the connection method needs to be at least as strong as the truss construction. Joints should therefore not be 'weaker' at all. The reason the truss appears to sag at those points is because there is often play in the connectors which when taken up under load makes it look like the truss is bending more at those points. Connections may also be stiffer than the truss section and therefore the stiffer bits don't bend like the rest of the truss. When the truss is levelled to take account of such anomalies, this can have a significant effect on the distribution of load on the suspension points. See above.

Load pins in a 'shear' connection require a retaining clip or equivalent to prevent the load pin from being drawn into the connector. It is alarming to see PVC tape (or even paper clips) being used to stop the pins falling out, in the belief that the load pins will be held in place by the tension or compression in the truss chords when loaded. No manufacturer spends money on drilling holes for retaining clips if there is no structural need. Check the manufacture's instructions on the correct type and method of connection.

## Slinging

Is it necessary to sling all four chords of a truss? That the method spreads load over the four chords to reduce the shear force locally is undoubtedly true, but usually unnecessary since the maximum allowable shear force in the material will be far in excess of the crushing force applied by the sling around the chord. The practice is probably a hangover from when riggers were understandably keen to use as many safeguards as they could since they felt 'the show must go on' even with no service history or certification on the equipment. If the integrity or load capacity of a truss is in doubt, don't hang it.

## Dynamic Loads

There are essentially two load conditions, static and dynamic. Static loads are loads that do not move. Dynamic loads are not considered as much as they should be. A typical chain motor lifts at around 4-5m per minute, producing a 'snatch' effect roughly equivalent to a 25% increase in load on the lifting equipment, the load itself and the supporting structure. This effect occurs both when hoists start and stop. Faster lifting speeds increase the effect. Manufacturers offer 'soft start/stop' machines to dampen this effect and modern automation systems generally 'ramp' up and down for the same reasons. Many trusses will exhibit a dynamic effect themselves, 'bouncing' when being lifted especially on long spans and when started and stopped. This itself may dampen some of the dynamic effect on the truss but increases load on the supporting equipment and structure. Always know the load to be lifted, check this against the manufacturer's load tables and that the dynamic effect is allowed for.

In summary, *know* how the truss was designed to be used – don't guess!

PPE Inspection Calendar
PPE
SHREDDER

# Rigging Call #24

*February 2009*

## PPE Standards

Many technicians in the entertainment industry are using Personal Protective Equipment* (PPE) that does not meet the standards required by the PPE Directive, implemented in the UK by the Personal Protective Equipment at Work Regulations 1992. An example of this is that many people are now aware of is using sport climbing harnesses for work activities. Less obviously, but no less unusual, is using a climbing 'tape' as a fall arrest lanyard. It is still common to see webbing slings choked around a harness ring and then choked to a scaffold hook as a rigger's means of fall protection. This practice means the strength of the webbing may be severely weakened and there is no energy absorbing element in the system. Often this sling will be attached to the central work positioning ring on the harness, not to a fall arrest attachment point.

It is also common to see two people sharing an anchor point or several riggers attached to a horizontal fall arrest system without knowing if the system is designed for multiple users. The steelwork or other means of anchoring must be reliable; if two people (or more) are attached to it, the forces involved may be significant if both people were to fall off. The use of climbing tapes as anchor slings is noteworthy.

The anchor for a fall arrest system should meet the test requirements of EN795. In the case of a sling, the standard is EN795 class B. The static test is to apply a 10kN force for three minutes to the anchor. The dynamic test specifies dropping a 100kg mass through 2.5m on a 2m length of three-strand nylon rope. Many technicians use slings which comply with EN566, a mountaineering standard. Whilst there is no doubt that the climbing sling may perform equally well in the event of a fall, the use of the appropriate standard seems to be a sensible approach and in the event of an accident means everyone is 'covered'.

Do we have to use standards when buying, selecting or specifying PPE for work at height?

Well, yes we should. The PPE Directive (and thus the PPE at Work

---

* Although the UK requirements for provision, storage and inspection of PPE are common to most kinds of PPE, references in this article are to PPE for work at height.

Regulations 1992 in the UK) defines 'complex' PPE as that designed to protect the wearer from 'mortal danger', such as falls from height, and that as complex PPE, it must be tested by a 'notified body' against the relevant standards (e.g. EN 361 for fall arrest harnesses). In simple terms, the only way a manufacturer of PPE for work at height can market the product is to mark it with the relevant EN number (e.g. EN 361) and the CE mark plus the four-digit serial number of the notified body concerned, (e.g. CE 1234). To find this, look no further than Regulation 4 (2)(e) of the PPE Regs 1992 which states PPE is only suitable if it complies with any legislation which implements (in Great Britain) any provision on design or manufacture with respect to health or safety in any relevant EC directive listed in Schedule 1 which is applicable to that item of personal protective equipment.

OK, so far so good, but what is this 'Schedule 1'? It can be found in Annex 2 of the PPE Directive (89/686/EEC) which says in section 3.1.2.2 'Prevention of falls from a height', that PPE designed to prevent falls from a height must incorporate 'a body harness and an attachment system which can be connected to a reliable anchorage point'. It must be designed so that in foreseeable use the vertical drop of the user is minimised to prevent collision with obstacles, and that the braking force does not attain 'the threshold value at which physical injury or the tearing or rupture of any PPE component which might cause the user to fall, can be expected to occur'.

This value is usually taken as 6kN in and as such is tested against the requirements of EN355. It must also ensure that after the fall has been arrested the user is maintained in a position in which he may await help if necessary. (Ideally sitting suspended from the sternal point, or less comfortably from the dorsal point). The Directive also requires manufacturers to supply all relevant information relating to the characteristics required for the reliable anchorage point (strength required, direction of force applied, etc), the minimum clearance necessary below the user (maximum lanyard deployment etc), the harness fitting instructions and how to connect the 'reliable' anchorage point to the harness via the appropriate lanyard and necessary connectors.

It could be argued that all this detail is overly prescriptive, but in the event of a fall of even quite modest proportions they could be critical factors.

PPE must be subject to regular inspection by a competent person. UK legislation apparently does not specify the frequency of inspection intervals, making it the responsibility of the employer or self-employed person to decide this based on the use they make of the PPE in question. BS EN 365 recommends PPE is inspected every 12 months but also that the manufacturer's advice should be observed. This means owing to heavy use or adverse conditions inspections are made much more regularly; three months may not be excessive. Inspections must be made by a competent person, that is to say someone who has enough practical and theoretical knowledge to be able to be certain the PPE is safe to use under the foreseeable conditions of their particular use. A user may not be competent to inspect equipment beyond a pre-use check.

A record should be kept of examinations for PPE used in work at height, including, and the details that need to be recorded are specified in the Work at Height Regulations. (Regulations 8(d), 12 and Schedule 5, parts 1-5, if you're interested).

TEST LABORATORY
35.40 Kn
TEST 1
MBL 5250 Kg
MBF 51.48 Kn
WLL 1000 kg
SWL 500 kg
8:1

# Rigging Call #25

*April 2009*

## Safe working loads

The terms Safe Working Load and safety factors seem to cause a great deal of confusion. So to begin with, here are the relevant abbreviations defined.

**SWL** = Safe working load
– the maximum load to be applied by law

**WLL** = Working load limit
– the designed load limit advised by manufacturers

**MBL** = Minimum breaking load
– obtained from testing samples to destruction

A Safe Working Load is the legal maximum load to be applied to a piece of lifting equipment. The SWL is specified by a competent person on behalf of an employer based on conditions of use, best practice and legal requirements. It should be noted that it is the responsibility of the employer to establish and mark safe working loads, not the manufacturer or supplier.

WLL is the manufacturer's recommended load limit and could be regarded as the same as SWL if the equipment is made for lifting and the safety factor is appropriate. A well known brand of shackle is marked with WLL not SWL. These shackles are made for several markets but are principally intended for lifting applications. The WLL allows an appropriate factor of safety for lifting, with 6:1 against the MBL in the case quoted, the load required for the shackle to break. On the other hand a manufacturer of dingy pulleys may quote a SWL with a factor of 2:1 against an MBL – meaning that if the quoted 'safe working load' is doubled, the pulley may fail. Still a 'safe' load – but in a situation where risk of injury is unlikely as a consequence of failure – it's not being used for overhead lifting. So what is defined as 'failure'? The item breaking? Starting to bend or stretch? Reaching its limit of elasticity (not returning to original dimensions when the load is removed)?

Lifting equipment is traditionally associated with a 'factor of safety against failure' meaning the point at which the bonds between the particles in an item bid farewell to each other. This is very different from, for example, loading figures quoted for channel framing systems used to support lighting bars. These figures relate to the allowable force on a given span against a specified deflection. The deflection will be well within the elastic limits of the material in question and thus able to return to its original shape when the load is removed. This expected deflection is normally given as a fraction, for example 1/360 of span, for example the load specified will create a deflection of 8.4mm over 3000mm.

This is not a traditional 'Safe Working Load' as used for lifting equipment. A safety factor is a ratio based on the failure load or other limit (such as deflection, as above). A shackle in the EC requires a factor of safety equal to at least 4 to meet the requirements of the Machinery Directive (Supply of Machinery (Safety) Regs in the UK). It is worth noting that BS 7905 and 7906 specify a factor of safety of 8 for the majority of equipment used over stages and similar areas. Every material can be overstressed to the point of failure; safety factors are one way to prevent those safe limits ever being reached.

A situation where this confusion is often encountered is when someone decides a certain structure or piece of equipment 'can't be used for lifting people because you need a 10:1 factor of safety' (sic). (Regulation 5 of LOLER requires lifting equipment for lifting people to have a factor of safety at least twice that normally used for lifting). So, you're lifting a 100kg person with a hoist displaying a SWL of 1000kg. That gives a 10:1 safety factor against the *SWL* of the hoist, not its MBL. The hoist already has a factor of safety of at least 6 on the chain, of around 3 on the brake and as much as 10 on the gearbox. Even with its overload clutch set at 1250kg there is still a safety factor in excess of 10.

However, important as the safety factor is, LOLER also requires that people being lifted aren't stranded through power failure, or likely to be crushed – the extra precautions that need to be taken when lifting human loads are not limited to increased factors of safety. Let's suppose the combined weight of two aerialists each weighing 70kg is to hang from one 2000kg roundsling. Let's also suppose that rigging for an aerialist qualifies as 'lifting' – that's another story; we'll assume for now that it is 'lifting'. Roundslings usually have a safety of 7, doubled therefore 14. To obtain a safety factor of 14 against failure, multiply (the aerialists) 140kg by 14, giving 1960kg as the MBL required to achieve 14:1. A roundsling with a SWL of 2000kg has an MBL of 14000kg. In other words the roundsling actually has a safety factor of 100:1 when used in this application.

Far more important, I would contend, is the structure supporting the equipment. The design loads for the building structure need to be known in order to determine what, if any, extra loads can be applied over and above the structural design. Buildings usually have roof structures designed to keep the weather out, not to support lighting rigs, PA systems or bungee ropes. The consequences of failure when the load is human or being lifted over humans need to be carefully assessed; increasing a safety factor is only one way of enhancing the security of performers. The job of the rigger has as much to do with establishing the loads to be applied to a structure as it does in making the attachments to it.

# Rigging Call #26

*July 2009*

## Knots

Knots are essential to many rigging tasks and often overlooked in this increasingly technological age. Using rope is something we take for granted and may dismiss as less important in contrast to other more sophisticated methods of lifting.

Most tasks on the technical side of the entertainment industry involve rope at some stage. This may range from tying the ladders to the roof rack on the van to rigging a show in a hemp house to a rope access task. Clearly the use of rope in theatres and other venues has declined in favour of wire rope winches and chain hoists but there are still few venues or shows that do not use fibre ropes in some shape or form.

Rope is generally supplied in coils, hanks or off a reel, that is to say there are no loops or eyes at the end. The user needs to be able to terminate the rope in order to use it for the job at hand. A splice, either mechanical or hand made is usually the strongest method of forming an eye, but this can be time consuming and requires a skill that the user may not have or need. The use of rope is often temporary and as such the cost of such labour intensive terminations as a splicing is out of proportion to the task. The work probably only requires a temporary termination and one that can be adjusted easily. So time after time, the user of a rope or cord will need to tie a knot.

Knot means a lump or tangle – as in a knot of wood, from Middle English *cnotta* and later *knutt* in old German. In fact a knot is very often not necessarily a 'knot'. There are three methods of terminating a rope, all commonly referred to as knots, irrespective of type. Even the use of the word rope to define a particular length of rope is technically incorrect – rope is the raw material as supplied in coils or hanks, etc. Once cut to a length and being given a purpose, some rope becomes a *line*. Knots are actually used to tie ends of ropes together, as in tying up a parcel. Bends join ropes to other ropes – as in a sheetbend, also the 'knot' used to make a net. Hitches secure a rope to something else. A common example in theatre would be the clove hitch, used to tie off a line to a flying bar, or the half hitch (or better yet, thumb knot) often used to secure the tail of a line to the standing part.

Knot strength is generally determined by the degree of stress the fibres in the rope experience in the knot form. In other words, the tighter the knot – the smaller the radius of bend in the fibres, the weaker the knot. So a simple overhand knot will be much weaker than a figure eight knot tied on the bight (tied on a doubled line to form a loop). Obviously the type of material used to make the rope also has an effect on knot strength. Some modern materials such as Kevlar are very strong if the fibres are kept straight, but not when bent over and they can be brittle compared to say nylon or hemp.

A strong knot, bend or hitch is therefore generally one that does not create severe bends in the rope fibres. A figure eight on the bight is therefore stronger than the popular bowline. However the practicality of the user being able to quickly untie a bowline after heavy loading is typically of more importance than its strength. The lines used in the entertainment industry are generally hugely strong in relation to the load they are lifting. The size of rope, usually identified by cross sectional diameter, is determined by the effort that can be applied in hauling a line that is light, strong and small enough to grip. If three people can haul 150kg they may be using a 16mm polyester rope, the breaking strength of which will be over 4000kg. When terminated using a clove hitch, this figure drops by perhaps 35%, still 17 times the load being lifted. If a failure was to occur, it is more likely the knot has been tied incorrectly or incompletely and becomes unfastened in use.

There are many hundreds of knots, bends and hitches, many with centuries of successful use. Knots are really a testament to human ingenuity and determination. The best are elegant and simple to tie. The real skill is knowing which to use in a certain application. The saying 'If you're not certain which knot, use a lot' is really anathema to anyone who knows about how to terminate a line. Sailors, fishermen, hauliers, stevedores, surgeons, lightermen, riggers, engineers and of course highwaymen have all contributed to the huge number of knots (bends and hitches) that exist.

The best knot will consider the rope construction, the strength required for the application, ease of tying and as importantly untying and the conditions in which the line will be used. If the line is going to be repeatedly loaded and unloaded, security is important but untying is not necessary, then a figure eight on the bight is often a good choice. A line that constantly needs an easily adjustable loop tying in the end and retying easily without worrying too much about rope strength will probably call for a bowline. Hanging a tube or barrel on a line or lines is often best achieved using a clove hitch with stopper or seizing, or a round turn and two half hitches.

So even in these days of intelligent hoists with variable speeds and zero fleet angles, being able to tie a good knot is as important as it ever was.

VENUE
MARSHAL

# Rigging Call #27

*September 2009*

### Whose venue is it anyway?

No-one wants to risk life and limb, but the industry we are in is geared to the show going up on time and on budget. From the producer to the local crew everyone is always mindful of the doors opening at 7pm. This is a huge pressure sometimes and we all know stories about the decisions that have been taken, rightly or wrongly, in order to get a show on. The art of compromise is one of the essential skills on the technical side of our industry, especially when touring shows. A significant number of present day managers are of a generation that learned by experience and current regulations may not be as well-known as they should be.

Generally if show goes up on time, things look OK and nothing bad happens the show is a technical success. This is no excuse, but I can sympathise with managers who have to balance the constraints of time and money against apparently ever more stringent health and safety requirements. However, the requirement in law means managers are engaged to represent employers and should know what regulations apply – working time, work at height, manual handling, lifting safety, electrical safety; the list can be seemingly endless. Some areas of production are sacrosanct aesthetically or from a safety point of view, but when knowledge or experience is lacking, corners are sometimes cut with the best intention but at the same time compromising safety.

Contravening fire safety precautions is perhaps the most obvious serious example, followed closely by manual handing issues and electrical safety, but often also structural or mechanical risks which include lifting, rigging and flying. Part of the responsibility of making sure shows are put on safely rests with the premises that house the production. A stationary venue is the easiest place to regulate, so if the production staff are lax the venue has the duty to ensure standards are maintained. Since it will also benefit financially from the production in most cases the venue has a joint duty of care over the activities in staging an incoming production and the personnel involved with it and an obligation as it is providing the premises and often at least some facilities.

Therefore venue managements have a duty of care over anyone on their premises, not just the public. If the visiting employees or contractors are not matching the standards required by the venue or licence conditions, the host venue steps in. An employer or self-employed person has a legal duty to ensure those doing the work have adequate knowledge and experience to carry it out safely and correctly. Legally, lifting operations (which is often what 'rigging' means to us) are assumed to be planned and carried out safely by competent people. The law requires it in any lifting activity; it's no different in the entertainment industry. The loads on lifting points need to be known and checked with the venue. The attachments may need to be made by appointed contractors rather than the visiting production. Attaching loads to lifting media is a skilled job as is the supervision of the lifting or lowering operation. Equipment must be fit for purpose and correctly used, meaning the venue facilities such as 'house' rigging points must be safe to use and the loading information required to use them safely must be provided. Likewise the incoming production must be able to prove their lifting equipment is safe to use and that their personnel are competent to work at height and carry out rigging work.

How does a venue ensure that rigging work being done in the venue is properly planned and carried out by people who are proficient? Historically, technical staff are (forgive me, everyone) basically enthusiasts. They learned what they know by experience, often literally by trial and error. Nothing wrong with that per se, but when something goes wrong, these days questions may be asked that they and therefore the management can't answer. Often it will be only the house electrician who has had any recognised qualification. That qualification is almost certainly related to permanent installation and although the principles of electrical safety and theory will be relevant, very often the qualification isn't directly relevant to the work being carried out on a stage. In the case of rigging as a trade (manual and mechanical handling, work at height) there has until very recently been no qualification relevant to the methods used in entertainment industry. Many productions and venues use freelance riggers or rigging companies to carry out rigging for them but both client and venue management are required to 'vet' their contractors' competence – either may be held responsible if things go wrong since they are employing those contractors.

Checking rigging competence is a difficult thing to do, especially if you're not a rigger. This is not to say that riggers without qualifications (and there are plenty) are either unsafe or incompetent, far from it. Clearly a person does not have to possess a formal qualification to be a competent rigger. However the best way to demonstrate competence is surely to be assessed by qualified assessors against national standards developed by the industry you work in. A nationally recognised qualification sets a benchmark for employers and venues so they can be assured that a rigger who holds the qualification meets the standards required to work safely and efficiently.

The National Rigging Certificate assesses a rigger's experience and the knowledge and skills that riggers use every day, no more, no less.

RIGGERS ON STAGE

# Rigging Call #28

*November 2009*

### Rigging is a word that means different things to different people

Needing a definition may seem pedantic but it is important for the industry to understand the term and what a rigger does, especially now the industry is taking notice. To most people 'rigging' means one of two things, the equipment used in hanging a structure up from which to suspend production requirements, or on the other hand it is construed as the hanging of luminaires (usually) on a supporting structure such as a lighting stand, barrel, scaffold or truss.

The roots of the word seem to date to the early 16$^{th}$ century and pertain mainly to sailing ships. 'Riggyng' or 'ragging' referred to the lines and chains used to support masts and booms and to control sails and anchors. The word can supposedly be traced back to Old Norse for 'to bind'. Subsequently 'rigging' referred to anything that required supporting, usually by materials in tension and often temporarily and also used to mean fitting anything out or to equip something. By the 1800s rigging also meant fixing a desired outcome in advance or to manipulate something dishonestly… 'Arriving at a desired outcome' is probably a reasonably apt analogy. As theatre technology developed, the ship-like arrangement of ropes and pulleys to hang or move scenery was also termed rigging.

The entertainment industry has started to use the term in a more definitive way. Rigging is in the main concerned with the rigging of powered lifting equipment for lighting, sound and video for shows or events in arenas and concert halls. It is often referred to as 'arena' rigging though the methods are becoming increasingly common in other areas. Rigging for exhibitions and corporate events also includes hanging drapes, banners and products as well as lighting and sound requirements, using both 'arena' and more traditional rigging methods. In the theatre, rigging is usually understood to mean rigging (hanging) lighting, sound or scenery onto an existing structure although it increasingly uses arena rigging methods in addition to the traditional means of flying.

In television studios, the activity is similar to the theatre, usually rigging lighting, sound, scenic elements or effects onto installed lighting hoists or scenery winches. There are also other discrete rigging categories, for example in the film industry where rigging is a hybrid incorporating scaffolding with rigging, often carried out specifically for special effects. Outside broadcasts use a mixture of film and television rigging methods. There is also specialised rigging for aerial work – essentially rigging for the suspension or movement of people, and the suspension or installation of works of art or exhibits in a museum or art gallery.

Work at height seems frequently to get confused with rigging. Anyone in a harness is understood to be 'a rigger' or to be 'rigging'. Anyone undertaking rigging is assumed to need a harness. Riggers may well need PPE, but so do lampies, chippies and noise boys; it's not necessarily connected with rigging. To those involved with the National Rigging Certificate, rigging is defined as 'the installation, removal or other activity using lifting or suspension equipment or accessories, used in tension for lifting or supporting display, production, performance or event technical requirements'.

When the various styles of rigging are examined it is clear there are many common features. These days using inverted chain hoists and trusses is common to theatres, studios and arenas. The use of steel wire rope is almost universal. Using ropes and pulleys, scaffold tubes and fittings are commonplace and in all branches of rigging working at height using access equipment is usually a factor. There are many similarities between the various branches of rigging, any differences are in the mix of skills required by each job.

One of the noticeable changes in recent years is that riggers are being hired (but by no means always) to carry out the installation and operation of mechanical handling equipment rather than letting the lighting/sound/stage department do their own rigging. It is slowly becoming accepted that even on small jobs, a competent rigger is a minimum requirement to ensure the safety of the lifting and rigging work and of the production that later relies on it.

So how to clear up the confusion? Employers and educators must understand the nature of the rigging in question to ensure that they are referring to it correctly. Using the right language is essential in this regard; in the theatre in particular it is very important. For example, to hold a qualification that refers to rigging could have disastrous consequences if the competence to rig hoists for a lighting truss is what is actually required of the technician, or in film, the rigging in question was understood to mean scaffolding and an arena rigger was contracted to carry it out.

In each sector 'riggers' should be the people responsible for the lifting or suspension equipment. Other rigging should be qualified by description – 'rigging lighting' and so on. Rigging lighting equipment in a theatre or studio is a necessary part of a lighting technician's skill-set but needs to be referred to as such rather than just as 'rigging'.

---

*Footnote: Readers may not know that Harry Donovan passed away in September. Harry was helpful and kind to me on every occasion I can remember; and he was very complimentary about my own offerings. His book 'Entertainment Rigging', (formerly 'Arena Rigging') is without doubt the best book on the subject and will continue to be of great help to riggers worldwide.*

# Rigging Call #29

*January 2010*

### Inspection of PPE of for Work at Height

The legal requirement to inspect PPE for work at height seems to be something many people have been unaware of until relatively recently. It seems as though a cursory glance and filling out the record card supplied with their harness was all that was believed to be required by many technicians, and more worryingly, their employers.

The term employer includes the self-employed. Often freelances assume themselves to be self-employed because they are invoicing their client. From a taxation point of view they may be self-employed, but very often will be treated as employees under health and safety legislation. Freelance workers in the entertainment industry often provide their own PPE for work at height. In many cases this means not only is the employer is not controlling the type of PPE being used but also that an employer is allowing PPE to be used uninspected. The client may legally be the employer and unless arrangements are made to the contrary could be responsible for issuing appropriate PPE, training employees in its storage and use, and ensuring inspections are carried out.

Employers or self-employed people must be able to demonstrate PPE for work at height has been inspected as required by legislation (PUWER, LOLER, the PPE and Work at Height Regulations) by having a current record of inspection. This record should be available to the enforcing authorities if requested and when equipment is used 'outside the employer's undertaking'. If you hire or borrow PPE, the record supplied should assure you of its safe condition in the same way as hiring lifting or other work equipment.

In terms of frequency, the law requires PPE to be inspected at least annually by a competent person on behalf of the employer or self-employed person. Although the PPE Regulations require fall arrest PPE to be inspected at least every 12 months, British Standards, industry guidance and sheer common sense recommend more frequent inspections for PPE subjected to arduous conditions of use. A lanyard in daily use, for example, may require inspection at weekly or monthly intervals.

The requirement to inspect work equipment for work at height that is exposed to deleterious conditions comes from the PUWER Regulations. The Work at Height Regulations further require employers to record inspections made of work equipment used for work at height, when and by whom and that the equipment is safe to use.

Inspection requirements in modern legislation are generally risk based on rather than by calendar intervals. PPE for work at height may include equipment used for work positioning or rope access, and is therefore not classed as PPE. Equipment used for access and rescue can be defined as 'lifting equipment for lifting people' under LOLER, requiring either six-monthly inspections or at intervals in accordance with a written inspection scheme drawn up by a competent person. Helpfully, kit inspected under LOLER satisfies the inspection requirements of the Work at Height Regulations.

Are freelance riggers competent to inspect their PPE? Some PPE may require specific knowledge or testing in order to be sure it is safe to use. Would they have access to the information required to be sure that an item has not been worn or stretched beyond permitted limits? Items like retractable lanyards may need specialist attention.

If there is a record of inspection at all, is it legally compliant? This is not to say the freelance can't inspect their PPE. Written information and specifications from a manufacturer or supplier or via a training course may bridge the gap between incompetence and competence. If knowledge is kept up to date and regular inspections keep skills practiced there is no reason why the freelance can't be competent, but impartiality is essential. For example, the cost of PPE is often the subject of complaint amongst the freelance community and a damaged £200 harness may be unlikely to fail at inspection for some time. Using a third party to inspect PPE ensures impartial inspection and correct recording methods. Though an external body will charge for its services, it is an expense that can be offset against tax. Users can be trained to spot potential defects or wear and refer anything they are unsure of further up the management chain. The employer can train suitable staff to the appropriate level of competence or use external contractors.

How can you ensure that your PPE is really safe to use and stays safe to use, rather than having a quick look at it before you use it? Get information on what the law requires, get training in how to inspect the type of PPE you use (rather than for the types of PPE you don't use).

All manufacturers of PPE have to supply information sufficient to allow safe use of their products, and who better than a manufacturer to provide the most authoritative source of information on a product. Distributors frequently offer training in inspection of the PPE brands they represent. Hopefully an increased awareness of inspection requirements will also lead to an improvement in the standard of PPE used for work at height in the industry. It is still common to see 'sport' climbing equipment being used and very often no record of inspection is available for all of a freelance rigger's PPE. Inspecting PPE used for work at height is a straightforward but very necessary part of safe working at height. After all, it is usually there to save your life…

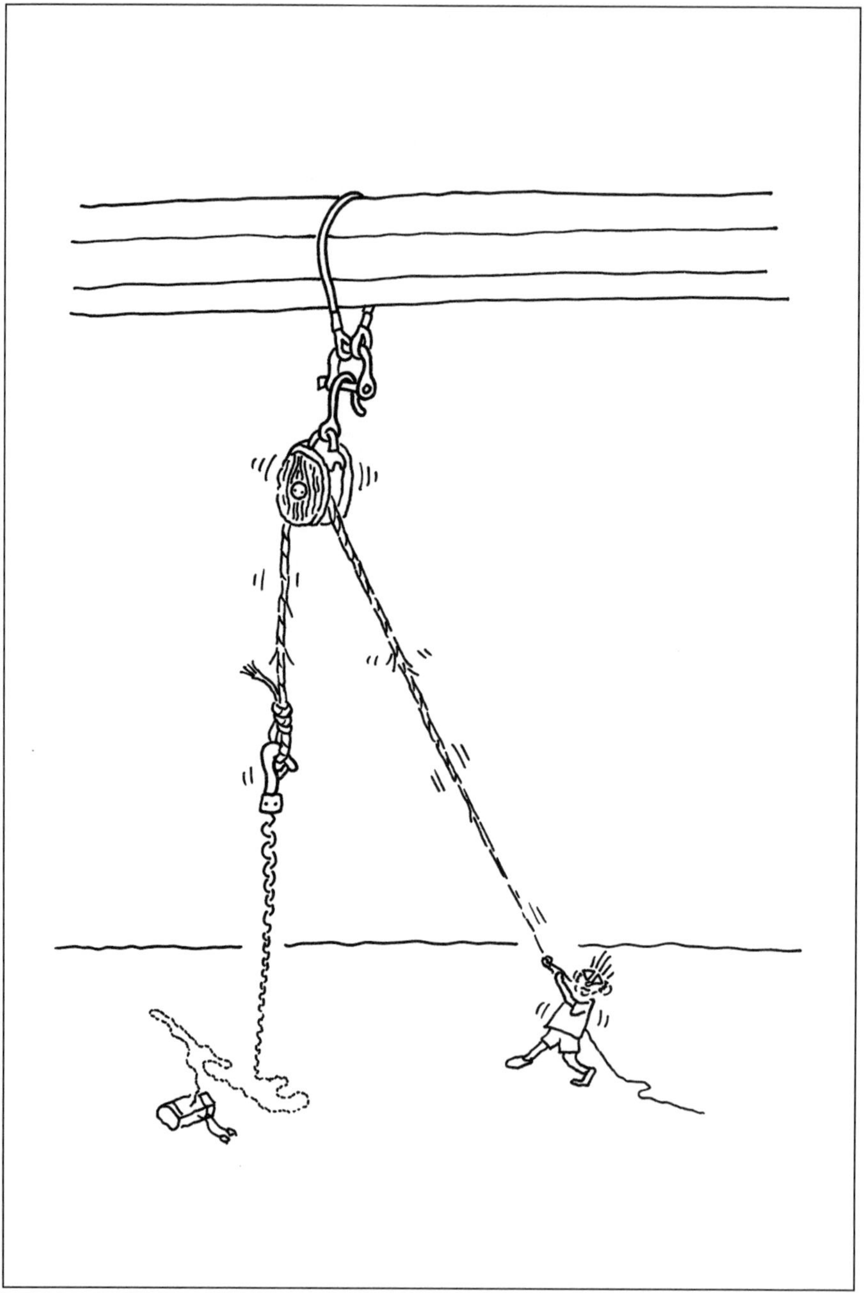

# Rigging Call #30

*March 2010*

## Hauling lines

Something that should have occurred to many riggers and flymen is that the humble workhorse of so many rigging operations, the hauling line and pulley, ought to be subject to exactly the same requirements as other lifting equipment. The line is used 'only' to rig or de-rig equipment but the consequences of a failure could be as serious as the failure of any other overhead suspension.

Most people would rightly regard a hemp set or spot line in a theatre as lifting equipment and expect it to be regularly inspected, have an identification number to avoid it being confused with other lines of identical appearance, marked with load information, and so on. So why not a hauling line and pulley? OK, it may usually be in a rigger's bag or coiled up on a fly-floor much of the time but it should be treated in the same way as a spot-line or a gin wheel and rope on a scaffold. What could go wrong? Well, the rope or pulley may fail through wear, possibly through damage caused by heat, friction or even chemical attack. Leaving a knot in one end will affect the strength of the rope over time and subject the same parts of rope to constant abrasion.

The rope used will often be polyamide (nylon) semi-static line of perhaps 11mm diameter with a breaking load of say 2700kg. With a bowline tied at one end and a factor of safety of 7, a safe working load of around 250kg is obtained when the line is new; much more than three sturdy crew members can apply to a thin hauling line. Using the pulley with a hauling line to raise a two fall hoist chain in a tall building can impose significant loads on the pulley block. A pulley carrying a hauling line up from the ground and back down to the ground (a 'single whip') will carry a load equal to at least double that of the load being lifted. A single sheave pulley is rated to lift a load of X. The load on the pulley will therefore be 2X because to lift X you have to apply a force of X on the other side of the pulley.

If the factors of safety against failure specified for lifting equipment in the Machinery Directive (thus in the UK the SoMR*) are applied, the

---

* *The Supply of Machinery (Safety) Regulations 2008 Schedule 2, Regulations 4.1.2.5 c), d), and Regulation 4.1.3.*

pulley and karabiner should have a factor of safety of at least 4:1, the rope and sling at least 7:1. A 30m two fall motor chain might weigh 80kg. The pulley therefore carries at least 160kg, allowing an extra 10% to account for the dynamic effect created by our three brawny assistants and a bit more for frictional losses the pulley and its attachment experiences a load of 180kg. If a factor of safety of 4 is used the pulley in this example therefore requires a minimum *breaking* load of at least 720kg though the load being lifted is only 80kg.

Usually a sling (not 'strop' – unless it has an eye at each end) and one or more karabiners attach the pulley to a supporting structure. The karabiner will have a minimum breaking load of around 25kN (2500kg) when loaded correctly, so divide by 4 to obtain a safe working load of 625kg. The sling will often be a 'climbing tape' (sewn sling) which when new might have a minimum breaking load of between 22 and 30kN. Applying a factor of safety of 7, the safe working load when used in a basket hitch over a smooth edged support with a thickness of at least 12mm will be compatible with the karabiner.

So in terms of strength, in good condition all the gear is probably fit for purpose. Many riggers' ropes and slings look like they have seen better days, the pulleys are unmarked and there may be no serial numbers and therefore it is impossible to record service history. Slings and karabiners are often unmarked and in poor condition, deemed too old to be used as PPE and relegated to use with hauling lines. To make sure equipment is in serviceable condition you must inspect it. This should naturally take place before each use but when treated as lifting equipment, it requires a more detailed and recorded inspection at intervals dictated by frequency and conditions of use. You could opt for the traditional six-monthly calendar interval, but given the potential consequences of a failure, inspection needs very much to be risk based rather than by calendar intervals.

It should be possible to produce a record of inspection for each item of equipment, allowing unique identification, a description and the safe working load of each item. This also acts as a reminder when calculating loads in various configurations. Keep the certificate of conformity that came with each item and mark the rope, karabiner(s), sheave and slings with a unique mark if they don't have one so they can be cross checked at inspection time. Marking must not alter the strength of the material.

All this can also be applied to items such as Klein tools, chain grabs and their associated karabiners or connectors used with the hauling line. All easy to do, making the working environment safer and helping you cover your rear should things go wrong…

FACTOR OF SAFETY? 2:1... 3:1?
70:1

# Rigging Call #31

*May 2010*

### Factors of safety

What is a factor of safety? What constitutes an acceptable safety factor?

If an item is designed to carry a load, its maximum capacity is usually quoted as a design load (sometimes called a working load limit, or WLL) so users are aware of its limitations and have a figure to refer to. The ratio between this design load and the ultimate tensile stress of the material is the so-called factor of safety. A 'Safe Working Load' (SWL) is subtly different in that an employer is required to establish an appropriate safe working load for the situation in hand. The manufacturer of an item may specify a design load or a SWL according to Standards or legislation applicable to the market sector, but it is the responsibility of an employer to ensure this ratio is appropriate when the equipment is put into service.

This is why, for example, a shackle intended by the manufacturer for use in lifting will already have an SWL obtained from perhaps a 6:1 safety factor against failure whereas a shackle from a ship's chandler may have the same WLL but based on a 2:1 factor of safety. Their intended uses are markedly different and the risk to people should the equipment fail is equally dissimilar. Generally, a factor of safety needs to be higher where there is uncertainty over the conditions of use or material properties of an item. This can be the case where the yield strength of a material (e.g. the minimum breaking load) is not definite or perhaps only specified within a range. The aim is to arrive at a design load where the elastic limit of the material in question is never reached, and foreseeable circumstances or conditions will not create weaknesses, assuring the user that the material will remain within safe limits and return to its original condition when the load is removed so it can be used again and again. If load and wear can be accurately predicted, lower factors can be used.

Deterioration due to the operating environment leading to unpredictable failure is often catered for by increasing safety factors. If the item is regularly inspected, allowing the situation to be monitored, the safety factor can be lowered in compensation.

Flexible material or things that get damaged in transit, being bent, knotted or dragged across abrasive surfaces need a larger factor of safety to account for this potential weakening by loss of or damage to the material. If the consequences of a failure are likely to result in serious injury or worse, for example where equipment lifts or is used above people, the normally used safety factor is increased, often doubled, to guarantee safety. A rope used to sustain a load overhead in a theatre must have a cast iron guarantee that it cannot break under *foreseeable* circumstances provided the design load is not exceeded. These foreseeable circumstances would include overload, environmental conditions, knotting and use round cleats or pulleys, but would not consider factors such as extreme heat or chemical attack because they are not likely to occur in normal day to day activity.

If the design load on a rope is 100kg, the safety factor needs to account for the dynamic increase in tension created by lifting (and lowering) which in a manual operation is unlikely to exceed perhaps 10%. However it needs to take account of any likely overload caused by, for example, the load being fouled – perhaps as much as 200% of the load being lifted. Already a safety factor of perhaps 3 seems prudent. Were an entirely predictable effect of sudden loading to include the rope being rubbed against an abrasive surface, there might be a significant loss of material, thus strength, so an allowance for this needs to be made.

Add to these factors the weakening effect of knots tightening, fibres being bent or crushed over time and it is no surprise therefore that the Machinery Directive requires a factor of safety of 7:1 for fibre ropes and slings. However, this performance would not be expected of a metallic item such as a hook, shackle or ring. Any loss of material will ultimately affect strength but in normal use they are unlikely to be affected by abrasion resulting in significant losses. Therefore a lower factor of safety is considered reasonable; the Machinery Directive specifies a minimum of 4:1. A wire rope is somewhere between the two and for general service 5:1 is the requirement. Legislation requires regular and timely inspection precisely for this reason, to allow defects to be detected before they begin to affect the factor of safety.

"What is the point of having a safety factor of 10 on the lifting equipment if the factor used in the truss only around 3?" Engineers designing structures are unlikely to consider 'failure' as meaning collapse. The safety factor used may be against an allowable deflection in the structure or failure characteristics of the material of connectors

| Equipment used in indoor venues | Material properties | Operating conditions | Loads and stresses | Proof loading | Regular inspection | Maintenance | Typical FoS |
|---|---|---|---|---|---|---|---|
| Trusses | Known materials with QA and certification | Predictable | Qualified design procedures | No | Yes | No | 3:1 |
| Metallic lifting accessories | | Predictable | Qualified design procedures | Possible | Yes | Yes | 4:1* |
| Hoists | | Predictable | High degree of certainty | Yes | Yes | Yes | 5:1* |
| Wire rope | | Predictable | Qualified design procedures | Possible | Yes | Yes | 5:1* |
| Fibre ropes and slings | | Predictable | Checked calculations used | No | Yes | No | 7:1* |

Factors of safety marked * are minimum requirements and must be at least doubled when used to lift people. Entertainment industry standards typically require an 8:1 factor of safety for overhead use of most lifting equipment and accessories.

used in the structure. This sometimes means a safety factor on a truss or structure may appear to be wildly at odds with other safety factors used in the lifting system. It is important to ensure equipment used in combination has compatible load capacity in the circumstances that are likely to exist and that safety factors are considered in the same way.

Where higher safety factors look desirable, perhaps a more thorough analysis of the issue should be carried out – they may not be the best solution.

MARKED
TO MARK

# Rigging Call # 32

*August 2010*

## Marking of lifting accessories

The method of marking small items of lifting equipment with necessary or legally required information has been an issue for many years. It is particularly relevant where thin wire ropes and small shackles are used. A search of the LOLER and PUWER ACoPs* seems to reveal no specific requirement to mark lifting accessories with a means of identification other than for the purposes of inspection records and where equipment may be used inadvertently to lift people. 'Schedule 1' of the LOLER ACoP requires a report of inspection to contain 'particulars sufficient to identify the lifting equipment including where known its date of manufacture'. Not quite the same as requiring a serial number, though common sense says a serial number is maybe the most sensible method. However it appears that legally (as opposed to the requirements of any Standard or Code of Practice) it is not a requirement to mark the item with a serial number. Interesting twist... I imagine it is something that insurance companies have required in order to offer cover.

The Supply of Machinery (Safety) Regulations 2008 (SoM(S)R 2008), which implements in the UK the most recent version of the EU Machinery Directive, states only that lifting accessories must show the type of material used where this information is needed for safe use and the maximum working load. In the case of lifting accessories on which marking the above is physically impossible, the 2008 Regulations state that the information required must be displayed by 'equivalent means' (plate, label, etc) and securely fixed to the accessory. Information must be legible not liable to 'disappear as a result of wear' or 'jeopardise the strength of the accessory'.

Clearly much, if not all, of this information is completely impossible to mark directly on a small diameter wire rope sling or shackle, but possible on a roundsling or metal item of reasonable size. The guidance given about marking SWL in the LOLER ACoP is that 'Where this is not possible, a coding system should be used...' Examples of such systems include

---

* The Approved Code of Practice for the Provision and Use of Work Equipment 1998 and for the Lifting Operations and Lifting Equipment Regulations 1998

colour coding or attaching some form of label. With regard to lifting accessories in particular, the LOLER ACoP advises if the configuration of an accessory affects the SWL, 'it should be clearly marked by a tag or a plate, or a chart should be readily available providing the user with information on the SWL for each configuration'. This suggests to me that the difficulty in marking essential information is understood and sensible alternatives are acceptable.

'Lifting accessories' are now better defined in the new Machinery Directive (and thus the SoM(S)R 2008) as 'components or equipment (including slings and their components) which are not attached to lifting machinery, allow a load to be held, are placed between the machinery and the load or on the load itself, or are intended to constitute an integral part of the load; and are independently placed on the market'.

("Lifting equipment" as defined by LOLER is work equipment for lifting or lowering loads and includes its attachments used for anchoring, fixing or supporting it.)

Assuming we want to use best practice and mark every similar item there are a few options to explore. OK, so a serial number isn't actually legally required, but common sense says you need a way of knowing, and giving each item a number is the easiest method. As you work your way through a large number of items you can tell which are yours, if you have inspected them previously, and with reference to records, when you put them into service. The manufacturer's batch number (if there is one) probably relates to several thousand shackles in the sizes likely to be encountered in the entertainment industry, so extra numbers will be needed if a unique number is to be created. First, the easy one – get the supplier to mark the items. This is typically by 'hard stamping' but this can cause indentations likely to raise stresses in the material.

Because shackles, masterlinks and ferrules have curved surfaces the depth of the stamping may be greater than the allowable tolerance for loss of material quoted by the manufacturer. This is particularly the case with items of circular section where in order to be legible the depth of stamping at the centre ends up very deep. If the manufacturer advises the item is too small to be 'hard stamped' (usually less than 13mm diameter material in the case of shackles and masterlinks, maybe 8mm for wire rope sling ferrules) then other methods should be considered.

If stamping isn't an option, tags that can be securely fixed and that will not get in the way or cut fibre slings can be used. Plastic 'write on' tags are another option as are pre-printed number tags. Some theatres manage very well by marking up each drift set with a tag detailing the show, its use and its SWL. When tags are visually obtrusive, once the drift is rigged the tags are removed and kept safe until the drifts are de-rigged, at which time they get re-attached. Markings should be legible, durable, consistent in convention and unambiguous. At least one manufacturer now supplies products such as shackles, eyebolts and hooks with a raised pad for the user to stamp their own identification, avoiding the depth of stamping issue. Some larger users etch directly onto their shackles and ferrules. This becomes cost effective for a large company with thousands of items to mark, but many smaller users will find it more efficient to order items ready marked in some way. The essential requirement is simply that the user has enough information to use the equipment correctly, and that each can be reliably identified.

NPR 8020-10:2006
BS7906-1:2005
REGLEMENT
ERP
BGV-C1
SS 767 1501-2
SQ-P2:2010

# Rigging Call #33

*November 2010*

## The PLASA International Rigging Conference 2010

The first conference of its kind in the world was held at Earls Court Conference Centre during the PLASA Show this year. The two-day conference was aimed at anyone involved with rigging in the entertainment industry.

The proceedings opened with a welcome from Matthew Griffiths of PLASA and an opening address by Alan Jacobi, managing director of Unusual Rigging.

Alan recounted the story of rigging in the UK accompanied by photographs and with anecdotes from well over 30 years of rigging for theatre, exhibitions, events and concerts: a thought provoking speech that recognised the fact that riggers are essentially problem solvers, people who are motivated by overcoming the engineering challenges presented in creating shows.

The second session was delivered by Nicky Greet, PLASA's head of skills and development who explained the development of the National Rigging Certificate in the UK. Nicky then posed the question about the viability or even desirability of an international rigging qualification. Riggers have been discussing this matter for years and the consensus of the conference was that it would be worth further investigation. Rigging methods are similar throughout the world today and to establish parity at least between existing qualifications seems a no brainer.

Murray Torrible took the lectern next, explaining the small print in our insurance policies as contractors, employers and freelance riggers. It seems it is strongly recommended we should all discuss with our brokers the scope of cover to establish exactly what we need and that they really do understand what our business activities are. It seems a policy typically offered as standard cover is usually nowhere near appropriate for many businesses in our sector.

Roger Barrett from Star Events was the next speaker; his topic centred on the advantages mother grids may offer and how alternatives to aluminium might be used in their construction. This proved a contentious issue; however it was a valuable contribution, though possibly not in the way anticipated. John Jones from Lift Turn Move delivered a short adjunct about hoists that conform to higher standards than typical industrial hoists and how they contribute further to the safety of personnel when used to list and suspend structures overhead.

Then came Mario. We'll be seeing more of Mario, I suspect. Mario was introduced as an Italian rigger embarking on a European tour who encounters different regulations in each of the countries he visits. Using cartoons to illustrate Mario's travels, Cristiano Giavedoni of CM Europe hilariously explored the requirements for overhead lifting safety, and the frequently discussed need for 'safeties' or secondary suspensions. Cristiano left me with the impression that a Europe-wide regulation isn't so hard to imagine…

A discrete demonstration in Earls Court 2 by Hawthorn Theatrical showing just how easily heat can affect rigging closed the first day. Polyester fibres will last only minutes under tension and exposed to relatively low temperatures. The sobering fact was that an aluminium ferrule secured eye termination (as found on most rigging 'steels') appears to lose its grip on the wire rope after about eight minutes at 250 degrees Centigrade…

Day two began with a change to the advertised programme because Stage One were delayed working on the Commonwealth Games. Stepping bravely into the breach, Kevin Crilly from the LEEA led a discussion about lifting accessories, specifically the need to mark shackles and other equipment. This included the need, or not, to mark shackles. On smaller items stamping can often indent material beyond allowable limits. Kevin also touched on 'the long link chain issue' but was unable to shed new light on the subject. Though the session created plenty of lively debate, no clear outcome was apparent – this one could run and run.

Conrad Schwarz followed with a session about the benefits of the host venue and contractor exchanging information. This was an interesting topic and one that had plenty of delegates thinking about something which is not immediately obvious: the legal duties each party has to the other, the financial relationship between them, and the information they should expect to offer or receive when taking a production into a venue. The outcome of such cooperation and coordination is a successful event.

Load monitoring was the focus of the next session. Eilon Engineering from Israel has been in the load measurement business for a long time.

Their ethos is that by knowing the weight of the load to be lifted and monitoring the load share we can make many lifting operations safer and more controlled – never more important than today where many shows involve structures weighing many tonnes being lifted, tracked or otherwise manipulated. Eilon's Ilan Bahar put forward an informative presentation and not in his native language.

After lunch we congregated in Earls Court 2 and were treated to a rescue demonstration presented by Harry Box of UK Rigging. We often think rescue from height needs to involve rope access intervention. If we follow best practice, we should be using powered access of some kind most of the time rather than scampering around trusses. Harry showed us that with a little planning and some basic training, it's possible to recover someone to a place of safety really very easily. The keyword? Planning.

The final presentation was by Mark Priestley of Unusual Rigging. Mark spoke passionately about how automation will require better training of a new kind of rigger, someone who understands the traditional methods of rigging but who can also operate automated systems that manipulate loads using complex sequences of cues. The fact that a great many shows now use automation as a matter of course should provide a wake-up call to us all that we need to recognise a new type of rigger.

The conference was closed by Alan Jacobi succinctly summarising the event and eliciting from delegates the desire to repeat the experience at PLASA 2011.

RIG OR DIE

# Rigging Call #34

*February 2011*

## Responsibilities and management

Sometimes it is all too easy to forget our responsibilities with regard to safety. We are busy, heads down, getting a show rigged, and not thinking about the consequences if things go wrong. If you are employed by someone as an employee, you are engaged to carry out work on their behalf. Therefore your employer is responsible for you and your responsibilities are defined by the terms of your contract and/or by legislation. Outside of direct employment, the responsibility for carrying out work more or less 'follows the invoice'. If you are an employee you are entirely 'owned'; as a contractor you are hired subject to a contract which is negotiated to identify responsibilities as well as scope of work and any commercial considerations.

So to cut a long story short, as an employee, you do what you are told to do. The method of work, planning, the equipment, workplace and supervision are provided by the employer. The employer takes overall responsibility for the work and if employing more than five people, should have written procedures or policy for what they employ people to do. Managers plan and organise supervision and answer to the employer or board of directors, for example. Supervisors implement the procedures drawn up by management on behalf of the employer. Employees carry the work out in line with instructions derived from the policies or procedures in place.

Very often it will only be when something goes wrong that a situation is examined closely – following an accident, for example. This is when written documents may be scrutinised and compared with what has taken place. For example, risk assessments may have been ignored, method statements not followed correctly, and defined systems of work circumnavigated because they take 'too long'. An employee's responsibility is limited by their position in the organisational hierarchy and job description.

So if you are responsible for doing a job in a particular way it is your responsibility to work the way you have been trained; the way the employer has decided it should be done. A supervisor ensures that you are doing so. An employee's role with regard to health and safety is essentially to look after themselves and anyone affected by their actions, in other words to maintain the safe conditions of work provided. These safe conditions are required of employers by law, but will also contribute to the effective execution of the work at hand.

So how best to ensure that employees take safety seriously and help themselves and the employer discharge their legal duties? It has long been the case that employers have had to provide 'adequate information and training' to employees to allow them to work safely and to reduce to an acceptable level the risk to themselves and anyone else affected by their actions. Employees need to know how to do their work, the hazards that may exist, and how the employer has decided to adequately control those hazards. In fact, what often happens is that employees make up methods of work as they go along. The employer's procedure may be perceived as too complicated, slow or onerous and a short cut is found.

In the entertainment industry this is a common problem; there are usually two parallel operations, the creative team and the technical team. The technical team is usually highly skilled and has a great deal of experience but they are not often consulted by management – especially with regard to scheduling work. Clearly the need to sell shows drives the enterprise but provided the show goes up on time without anyone or anything getting damaged in the process, technicians are often left to their own devices in carrying out work rather than in accordance with any procedures. This isn't to say technicians aren't being careful but their activities very often won't square with an employer's written procedure. This could well land the management in hot water should something go wrong. (This being the same management that scheduled the work in the first place....) So the issue is how to work safely, without sacrificing disproportionate amounts of time serving a well meant but burdensome health and safety policy.

To have a middle line between stringently prescriptive methods and 'making it up as we go along' is the key to this. If there are real hazards then simple and robust measures need to be put in place. Importantly these need to be implemented by supervisors and not written up and forgotten. Making sure the employee understands the employer's responsibilities to the employee and vice versa is a good place to start. Sessions to cover more specific tasks or techniques can be developed in this context, building on the understanding of the basics. To let the crew plough on, uninformed, unprepared and ill equipped seems to be condoned by some managements.

A small investment in people at the start of their employment and at regular intervals will reap benefits far in excess of the cost of their training. If everyone understands how the organisation is supposed to work, the hierarchy and lines of communication the organisation will be more efficient and safer as a workplace. Making this happen will always pose a problem to businesses – how to balance the time and cost required to provide adequate training. Training in-house makes it a living part of the business, keeps the information relevant, makes it cheaper, more effective and often of shorter duration – meaning it's far more attractive to the workforce. Having more than the absolute minimum of information is almost always a good thing, but taking it too far is counterproductive. The 'What's in it for me?' principle is generally the way we all learn the dull stuff and using this principle definitely helps instil an understanding of our responsibilities at work both as technicians and as management.

2000 Kg
LOAD MONITORING
545 Kg

# Rigging Call #35

*March 2011*

### Secondary suspensions, again…

Why do we need secondary suspensions on flown equipment such as trusses, PA or LED displays? Standards state a permanent installation should not normally need secondary suspensions because the design should engineer out risk. This is usually achieved by increasing safety factors or using 'redundancy' in the quantity of suspensions rendering secondary suspensions unnecessary. In temporary installations – for example touring shows, provided the same principles are applied I don't see why secondary suspensions are necessary except in specific situations.

In the past, production 'motors' were used sparingly because there weren't many available, so two motors per truss was often the rule. Trusses were often of unknown provenance and deflected alarmingly. So incoming equipment was secured to the roof as much as possible – just in case. Over the years, licensing requirements and advice from well-intentioned but sometimes misinformed local authorities has resulted in the house believing that 'safeties' are legally required, so they are dutifully insisted upon. Very often 'safeties' are installed badly – far too slack, incorrectly attached at either end and often with inadequate materials. So far from making the situation 'safer' they even create a false sense of security.

So when do we really need to consider using secondary suspensions? I would contend that if anything supporting a load could fail and that failure would result in the load becoming a hazard, something needs to be done, though that does not necessarily mean using a secondary suspension. Examples of this would be single and two-point suspensions. Another obvious situation is when there is a potential risk from fire or heat. Clearly, if a combustible material is being used, for example a polyester sling close to incandescent pyrotechnic fall-out, then some resistant material should be used instead. This may be a steel wire rope sling but it should be noted that if it has aluminium ferrule eye terminations the manufacturers' maximum operating temperature is often only 15 degrees Celsius higher than for a polyester sling. In the case of a fire, this may not offer the protection often assumed. ‘Soft’ steels may be a good alternative since there is no ferrule, the part most at risk of failure when heated. Another factor that might influence decisions with respect to fire is that an aluminium truss will lose up to 45% of its strength when heated to 200 degrees Celsius.

Single points are clearly a risk unless ‘over-engineered’ – in other words the safety factors are greatly exaggerated. Since lifting regulations in Europe at least, require a doubled safety factor when used to lift people, maybe a similar increase should be used when suspending loads over people. Many venues do not require the use of secondary suspensions where the load is less than 50% of the safe working load of the suspension for this reason. The safety factor is doubled. There are several standards in Europe that recommend this approach. The other sensible way to approach this is to increase the number of suspensions.

I recall an American show in the mid 80s using a ‘flip-box’ catwalk truss system with a lot (and I mean a lot) of hoists on it. If a point had failed, I doubt anyone would have even noticed. That is a very different picture from a two-motor front truss in a town hall where should one suspension fail the truss would swing or even fall onto the audience beneath.

So how should we install a secondary suspension? Assuming it is required at all, make sure it will work. A rigging ‘steel’ choked round one chord of a truss and wrapped/choked around a beam without softening at the other end is common practice. In the event of it being shock-loaded it is unlikely to support the truss adequately and if slack the load will accelerate creating a ‘shock load’ quite possibly beyond the capacity of the rigging or the supporting structure. The practice of using a chain clutch (grab) as a ‘motor bypass’ issue is a moot point. Though it can eliminate slack very effectively, the big manufacturers of hoists have said “don’t” because they are concerned about the security of the clutch on the chain or damage to the chain itself since the chain clutch is generally designed for the proportions of sling chain, not hoist chain. Statistically, accidents relating to hoists involve the chain/hook connections or load displacement, so the bypass method guards only against internal mechanical failure. Most hoist manufacturers say don’t leave suspended loads unsecured (or words to that effect) because of their obligations in putting their hoists on the market. LOLER in the UK specifically requires the use of additional precautions when suspending loads over people. I don't think that will ever mean just the audience or public.

Not using secondary suspensions also means the system can be quickly lowered in an emergency. It is important to recognise that people still climb on trusses. The wisdom of that is another topic for another time but we should consider being able to lower the system as part of an emergency procedure.

So, what is the solution? Every situation needs to be assessed for risk, but in general terms it seems sensible to say that when certain criteria are met secondary suspensions are not necessary. As a minimum, these would seem to be whcre:

- hoists are loaded to less than 50% of safe working load
- the load is properly supported by multiple suspensions from designated anchors
- truss spans are significantly reduced from the manufacturer's allowed maximum
- the suspensions are made in non-combustible materials appropriate to the situation
- in *all* cases *all* the equipment used has a current report of thorough examination,
- and crucially, that the failure of any suspension in a system would not result in significant
- movement of the suspended load.

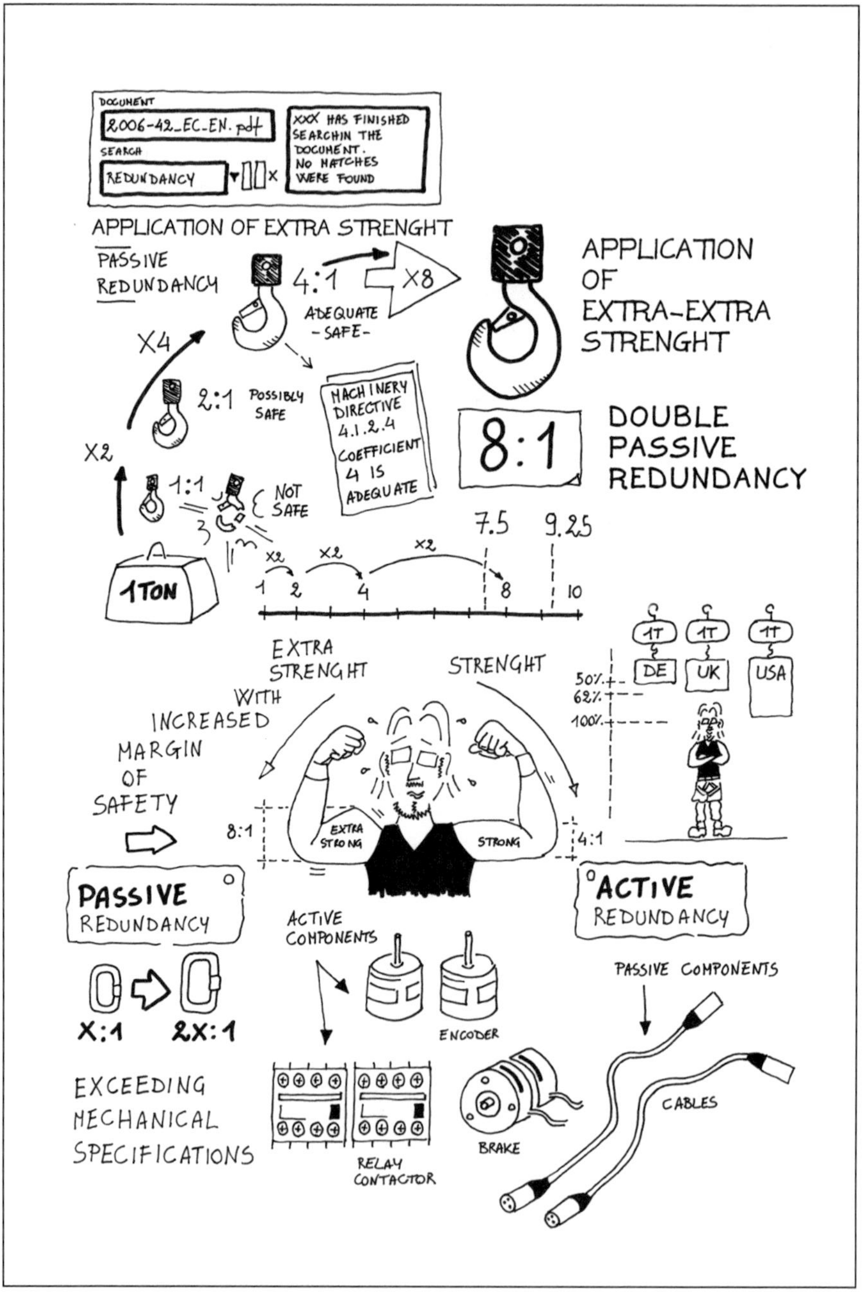
DOCUMENT
2006-42_EC-EN. pdf
SEARCH
REDUNDANCY
XXX HAS FINISHED SEARCHIN THE DOCUMENT. NO MATCHES WERE FOUND
APPLICATION OF EXTRA STRENGHT
PASSIVE REDUNDANCY
4:1
X8
ADEQUATE -SAFE-
APPLICATION OF EXTRA-EXTRA STRENGHT
X4
2:1
POSSIBLY SAFE
MACHINERY DIRECTIVE 4.1.2.4 COEFFICIENT 4 IS ADEQUATE
8:1
DOUBLE PASSIVE REDUNDANCY
X2
1:1
NOT SAFE
7.5
9.25
x2
x2
x2
1TON
1
2
4
8
10
1T
1T
1T
DE
UK
USA
EXTRA STRENGHT
STRENGHT
50%
62%
100%
WITH INCREASED MARGIN OF SAFETY
8:1
EXTRA STRONG
STRONG
4:1
PASSIVE REDUNDANCY
ACTIVE REDUNDANCY
ACTIVE COMPONENTS
ENCODER
PASSIVE COMPONENTS
X:1
2X:1
EXCEEDING MECHANICAL SPECIFICATIONS
RELAY CONTACTOR
BRAKE
CABLES

# Rigging Call #36

*May 2011*

## Mail bag

I receive a lot of emails asking for advice or at least some information and I thought it might be interesting to look back at nearly ten years of correspondence to see if there were any trends.

What I found was that the questions asked were split fairly evenly between rigging and work at height issues. The rigging ones are overwhelmingly to do with factors of safety and their relationship with safe working loads (or working load limits) and load testing. The use of man-made fibre slings is another common concern. The work at height stuff is often about the confusion between working at height with PPE and 'rope access', the perennial Tallescope issue and about fall arrest forces. The work at height questions I'll leave for another time, but I thought I'd re-visit a few of the rigging hardy annuals again.

A factor of safety (FoS) is the ratio between the load at which an item might fail and the load to be applied in use. 'Failure' does not always mean the item breaking – it could be a significant deflection or permanent stretching depending on the item in question. Legislation for lifting equipment in Europe requires a factor of safety for shackles, masterlinks, chain, chain slings and their fittings to be no less than 4:1, steel wire ropes and their terminations, no less than 5:1, and fibre ropes and slings, no less than 7:1. However, although not law, BS 7906 requires a factor of safety of 8:1 for lifting equipment used overhead in the entertainment industry.* The exceptions to this are bars, trusses or similar structures where something between 3:1 and 4:1 is more likely. A 'crude' ratio may be inappropriate to apply to these structures, so use the load information supplied by the manufacturer or installer.

## Working load limit or safe working load?

**Working load Limit (WLL)** is the manufacturer's maximum recommended load for a product.

---

* BS 7906-Part 1, Use of lifting equipment for performance, broadcast and similar applications. Code of practice for installation, use and removal of above stage equipment (excluding trusses and towers)

**Safe Working Load (SWL)** is a reduced load to take account of the circumstances of use.

This is decided by the employer's competent person, who is not necessarily the user. So to obtain an 8:1 ratio, take the marked SWL or WLL of the shackle – 2000kg. Multiply this value by the safety factor the manufacturer specifies to obtain the minimum breaking load (MBL) of the shackle; 6 in the case of a typical two-tonne alloy bow shackle used in entertainment rigging. So SWL of 2000kg x FoS 6 = 12000kg MBL. Then divide this MBL by 8 to obtain the 8:1 factor of safety required by the Standard. 12000kg / 8 = SWL of 1500kg. The maximum load you should now apply when using the shackle is now 1500kg instead of 2000kg to take account of the risk factors inherent in overhead use.

I would expect that in the vast majority of uses in the entertainment industry, a two-tonne shackle will usually achieve a factor of safety far in excess of 8 simply because it will be used not for its load capacity but its dimensional compatibility with other components in the load path. LOLER states that if working under suspended loads is necessary, the employer must take extra precautions, and list some options which include increasing safety factors. When lifting people the safety factor must be at least twice that used for conventional lifting; many standards suggest using the same ratio when lifting loads over people.

## Load Testing

It may not always be necessary to carry out a load test. A lot of the thorough examination required in law is visual and non-destructive. If you do apply a test load it shouldn't put the item anywhere near its breaking load. It is at the discretion of the 'competent person' to decide the nature of the test on behalf of the employer. Lifting equipment and lifting accessories must be capable of being tested to a specific value, generally as a static load test at 1.5 x SWL for manually operated hoists and lifting accessories and 1.25 x SWL for powered lifting equipment.*

## Roundslings

The current standard BS EN 1492-2 was published in 2000 and amended in 2008.

---

* Supply of machinery (Safety) Regulations 2008. Note it is a static load test, not dynamic.

The sling should bear a blue label (for polyester slings) showing its serial number, working load limit, CE mark and the configurations that affect the strength (mode factors). Usually the date of manufacture and safety factor will also be marked. The operating temperature for polyester slings should not exceed 85 degrees Celsius. Roundslings with cuts in the sheath exposing the white load-bearing fibres should be destroyed. Webbing slings have their load-bearing fibres exposed to ultraviolet radiation from the sun and will degrade over time. The sheath of a roundsling shields the load-bearing fibres from UV, so indoor use is unlikely to affect slings in the same way as use outdoors.

The practice of joining roundslings by 'choking' one to another is bad practice and potentially dangerous. The heat generated from friction as the 'knot' tightens can fuse fibres together creating brittle lumps. Roundslings are a continuous skein of fibres, the final turns of which are simply cut and taped to the skein. Consequently there are a number of pliable 'lumps' which can be felt through the sheath. If the lumps are hard or 'crunchy' suspect internal damage from friction and destroy the sling. As in all rigging matters, if there is any doubt in your mind about the strength of a material or structure you are working with, consult a competent person. This might be a qualified engineer, manufacturer or an experienced rigger. It should never be your best guess.

HELP ME...
...I'M SAFE!

# Rigging Call #37

*August 2011*

### Three points of interest and an observation

This year's PLASA Rigging Conference ($12^{th}$ and $13^{th}$ September) is being run along more or less the same lines as last year, but with fewer sessions, allowing more time for discussion and questions.

Steve Colley is opening the conference with the background to the House of Dancing Waters show and the challenges it presented. There is a strong international presence as there was last year, with Cristiano Giavedoni and Matthias Mueller speaking about standards and chain hoist design respectively. Chris Blakeley, a Brit now leading Petzl's V'Access research and training facility in France, is presenting a session on fall arrest forces. From the UK, Nippy Williams will be speaking about planning work at height and Ted Moore from Stage Technologies will be talking about performer flying and the challenges that produces. Panellists supporting the speakers are from the UK, Holland, the USA, Italy, Germany and New Zealand. The closing session will again be presided over by Alan Jacobi and will summarise the two days and chair questions from the floor on matters raised during the conference.

New for 2011, on Sunday afternoon $11^{th}$ September, a Riggers' Forum aimed at riggers who can't attend the main conference will be held. Delegates can expect a light lunch and a beer as well as the chance to discuss the issues that affect their working lives. The afternoon comprises two sessions. The first is a presentation by Michael Anderson of Earls Court about where responsibility for a job really does lie. The second session is an open forum chaired by Mark Surtees, chair of the National Rigging Advisory Group, and on the panel such stalwarts as Eric Porter, Oz Marsh, Sven Knight and others. The organisers are encouraging questions to be submitted in advance and there will be a prize awarded to each of the two people whose questions generate the most discussion. For more details see the conference website at www.plasariggingconference.com.

A small but notable victory for common sense was won in June this year. The ABTT and other industry groups finally reached agreement with the HSE about the movement of an 'occupied' Tallescope. Hurrah! For over ten years the argument has raged; now a solution is found.

You can move a technician in the basket **BUT** subject to the following conditions:

Section 5 of the soon to be re-published ABTT Code of Practice for selection of equipment for work at height in theatres must be followed and there should be an effective rescue plan in place.

The Tallescope itself must be modified as follows:

- braked wheels replace existing trailing castor wheels
- an extra set of outriggers is used (now two per side)
- 'push/pull' bars must be fitted to improve manual handling.

In my view the last modification is unnecessary, but a small sacrifice in order that Tallescopes can continue to be used as they have been for a long time and now even more safely.

(Note: Thankfully the 'interim measure' involving harnesses and clipping to an adjacent bar has been revoked. You don't need PPE when working in a Tallescope, it could even make it more dangerous unless carefully designed and always anchored overhead in such a way as to avoid destabilising the Tallescope. Astonishingly, there are still theatres where the house rule is that you use a twin fall arrest lanyard when climbing up or down a Tallescope. Just say no).

If the industry has learned one thing from this exercise it is that we need to make sure our methods of work are as safe as we can make them so we can tell the authorities how we do things before they tell us... Which leads me to this observation. Whilst the general drive towards a safe working environment is clearly to be applauded and supported, it has also led to virtually every contractor in the developed world wearing hi-vis jackets 24 hours a day. It seems to me we are slipping into a world of almost obsequious compliance. The whole point about hi-vis is to alert other people to your presence. The human eye is very quick to pick up movement and colour in peripheral vision. Accentuating our presence is a jolly good idea when marshalling vehicles or loading out across a busy street, particularly because it will often be dark...

What we don't need is management telling young people to wear hi-vis to the extent that they feel unprotected without it. They're thinking about avoiding the wrath of their supervisor rather than the hazard it is helping (and it is only helping) to protect them against. Wearing hi-vis merely aids conspicuousness.

Don't get me started on hard hats. When and where appropriate, of course we should wear them. Employers (including the large number of technicians in the industry who employ themselves) should conduct risk assessments and determine the specification of the PPE required after all other control measures have been applied. Get to the root of the problem. It doesn't mean wearing a hard hat simply because it's site uniform; wearing one means accepting the risk of being struck on the head by a falling object or swiped by a scaffold tube being carried by a colleague really cannot be controlled in any other way. What I'm trying to say is *please* can we return to a time when it was in the gift of the employee or at least supervisor to decide when PPE is needed rather than have a whole venue believing they are 'safe' because everyone is wearing hard hat, hi-vis and boots? That undoubtedly means educating the workforce in how to manage risk.

Surely employers (and their insurers) would much rather we avoid risk wherever we can rather than routinely fall back on using PPE which, as we are always reminded, should be the last resort? Rant over…

# Rigging Call #38

*November 2011*

## The 2011 PLASA Rigging Conference

Steve Colley and Abigail Matthews presented a fascinating talk about the 'House of Dancing Waters' at the City of Dreams resort in Macau. The sheer scale of the show, the numbers of cast and crew and enormity of the production requirements held the delegates in rapt silence for an hour. Quite an 'opener' and a fitting follow up to Alan Jacobi's inaugural presentation in 2010.

Cristiano Giavedoni gave us a considered view of the standards that exist across Europe. It seems as if we are reaching the point where common standards may be possible across Europe in terms of the expectations of venues at least. There are more similarities than differences and ultimately we all have the same goal – getting a show on safely. Ensuing discussions suggested the global rigging community needs to start getting together to establish common codes of practice in European venues at least. The compiling of a dictionary of rigging terms is a first step. Watch this space…

Matthew Griffiths, PLASA CEO, spoke about progress during 2011 with respect to the two certification schemes operated by PLASA (the ETCP programme in the US and the National Rigging Certificate in the UK) and how the needs of the territory concerned should determine the choice of certification or assessment. Each scheme has merit and Matthew mentioned that discussions must take place in order to identify the common ground.

Chris Blakeley from Petzl spoke with authority from both his own experience and of one of the biggest names in PPE for work at height. Chris's presentation was centred on fall arrest forces created by the 'larger individual'. Significant at the best of times, Chris was pointing out that these larger forces may be incompatible with anchor strengths typically specified for both vertical and horizontal fall arrest systems.

Steve Williams (Nippy) spoke about rescue and rope access. Steve's key points were that with a trained team, carefully planned work and appropriate rescue plans the likelihood of a rescue being needed at all is minimised. Another point well made was that having first aiders trained to work at height is often overlooked when training staff.

Ted Moore from Stage Technologies spoke with passion and in some cases nostalgia about the huge body of work that Stage Technologies has accomplished over almost two decades, notably the Millennium Dome show in London. It was clear from Ted's presentation that the 'art' in rigging is often best demonstrated in performer flying – a real synthesis of engineering, creativity and problem solving.

Technical consultant to the Lifting Engineers' Equipment Association, Derrick Bailes delivered a factual presentation developing topics from the 2010 Conference – notably the marking of shackles. The panel included Adrian Tinson from the HSE's entertainment and leisure section. The consensus was that provided an employer has properly managed method of identifying equipment as safe to use, there was not necessarily a need to mark every item with a serial number. Clearly this needs further discussion, but it was good to see the experts assembled broadly in agreement.

If you'd ever wondered about how chain hoists are designed, Liftket's Matthias Mueller delivered a detailed account of how the design and manufacture of chain hoists considers every last detail. We should be able to rely on the manufacturers to deliver a robust and reliable product and Matthias explained how Hoffman GmbH achieve this. Food for thought and confidence inspiring to know the lengths manufacturers go to. At the first Rigging Conference in 2010, there was a debate about the need for secondary suspensions and double brakes. This begged the question – do hoists fail often? Not an easy question to ask or answer in our industry. Ron Bonner, PLASA's technical resources manager made a determined attempt to do so during the year. After manufacturers declined to divulge details, Ron devised a questionnaire for users of chain motors. Over 130 responses, all be they anecdotal produced data upon which some conclusions may be drawn. Most notably, hoists failures are rare and provided they are rigged, operated and maintained correctly offer a reliable and safe method of lifting and suspending production requirements. The full presentation can be seen on the PLASA website.

Bob Stagg is not a name many people in the conference room were familiar with. In fact Bob has been quietly responsible for engineering solutions in the Albert Hall and other venues, notably for Cirque de Soleil, for many years. Clearly the voice of great experience Bob's often humorous commentary about finding solutions to the challenges associated with putting big shows into a building opened in 1871. As a complete aside, it was pleasing to hear Bob shares my views about the use of PPE and how site requirements often miss the fundamentals of managing risk.

Alan Jacobi was greatly missed at this year's conference; he had picked up a nasty bug in the Middle East earlier in the year. His place chairing the lively closing question and answer session was aptly taken by Mark Surtees of Outback Rigging, chair of the National Rigging Advisory Group in the UK. The rigging community expressed their wish to repeat the experience in 2012 and their thanks to the PLASA Events team for again producing such a worthwhile event so well.

This year there was also a Riggers' Forum on the Sunday afternoon preceding the main conference. The aim of this event was to allow riggers unable to attend the two weekdays the chance to meet and discuss matters of the day with their peers.

A healthy turnout heard an effective presentation from Michael Anderson (Group Head of health and safety at Earls Court and Olympia) speaking in plain English about who would actually be in charge when riggers are at work. This is the kind of information everyone working in the industry should have explained to them upfront.

The rest of the afternoon was given over to discussions chaired by Mark Surtees which ranged from certification to secondary suspensions – a worthwhile exercise and like the main conference one that will again be held in 2012.

# Rigging Call # 39

*January 2012*

## Risk Assessment

There can be few words in the English language more likely to cause eyes to roll and arms to be fold than 'risk assessment'. Since the introduction of the 'Six Pack' of European derived regulations in 1992 (Management Regs, PUWER, LOLER, etc) an employer or self-employed person has had duties to carry out risk assessments of their work activities. This will range from the generic for a particular activity such as manual handling or work at height to task-specific risk assessments relevant to one effect or one production. There may also be the more onerous 'dynamic' risk assessment. This option may appear to be the solution for people who want to make it up as they go along, but in our environment is probably only appropriate for activities of very short duration, or (where the method came from) in an emergency situation.

The method relies on close supervision and highly trained people since the situation being assessed in unlikely to be a safe one. There is still need for a paper trail too. Not the easy option it may appear to be, in fact. So why has the entertainment industry been so reluctant to see the value in risk assessment? Rigging is surely of all trades one of the most high risk jobs. So why is it when you ask a rigger for a risk assessment you get back a document which at best is a description of what PPE they intend to use, or how large an area they intend to cordon off whilst they work at height because someone might drop something?

How many times have we heard that there's no need for a hard hat? "If there are riggers up there who drop stuff, they shouldn't be riggers". Often venues insist the whole venue floor is to be a hard hat zone during technical work. This in my view gives a false sense of security – it might be a reasonable control measure in some cases but it doesn't address the cause; it may not actually even offer effective protection either. Somewhere between the two is where we need to be, sensible controls for real hazards. The standard acronym in safety circles is ERIC* meaning to consider as part of a control hierarchy each of these steps in order:

---

* (*sometimes 'ERIC PD', where P is PPE and D is discipline. P and D are individual measures. Priority is always given to 'collective' measures, those that apply to the largest number of people.)

Eliminate – do the work another way, one that entails no risk; if you can't then; Reduce – exposure to a hazard; if you can't then; Isolate – exclude as many people as possible from the danger areas; if you can't then Control – for example by using competent people and work to a method statement.

We all consider risk to some degree every time we sip a hot drink or go for a swim, hopefully in the same way at work before climbing a ladder or making an electrical connection. However part of the requirement at work is that the employer/self-employed person needs to be able to prove they have done so and are effectively controlling the significant risks they've identified.

The sticking point is rarely the ability to carry out the assessment; it is committing it to paper and then acting on the findings. So how do we make it easier? Our friends at the HSE have long recommended the 'five steps to risk assessment'. To summarise, it's about identifying hazards, considering who the hazards affect and what 'control measures' can be put in place to manage those hazards. The control measures need to be in proportion to the risk and to be included as part of routine procedure. Remaining risks are then quantified and further action taken to reduce the risk to the lowest acceptable level.

HSE advice states that 'risks from everyday life' need not be considered unless the work activity increases the risk. This is not a licence to ignore the bad practices that are often tolerated in the name of expeditiousness. It requires us to decide what constitutes significant risk, and the control measures required are effective and actually implemented. The outcomes of risk assessments in terms of methods of work, types of PPE and supervision levels can have a number of useful functions. They can help with induction and further training and will link with method statements and lifting plans. Probably most importantly they can be summarised and given to supervisors or team leaders to support them in directing and supervising work.

The recent Lofstedt Report recommends that reducing requirements on business that do not lead to improved health and safety outcomes, or that go beyond what the regulations require, will enable businesses to 'reclaim ownership of the management of health and safety'. To me that means discouraging reliance on well-intentioned but over prescriptive site rules or using health and safety consultants for the mundane or basic management of safety and health. We should be able to manage our own risks if we know what we're doing. So removing the pressure to produce reams of paperwork to allow someone to use a ladder might allow managers and supervisors more time to actually supervise the safe use of that ladder. The paperwork exercise should not be prepared just as a possible defence in case something goes wrong or instead of direct communication with people who do the work. Currently many risk assessments are carried out by people who don't do the work. The people doing the work will usually be the best the equipped to understand the risks inherent in a task. What people need is the ownership of the methodology, help with is the recording of the process, and the trust of management.

...NOT SAFE

# Rigging Call #40

*March 2012*

## Work at height – Liabilities and Responsibilities

In the space of a week I've received three enquiries which make me wonder a little if the Work at Height Regulations have had the effect they were intended to. Professor Löftstedt recommends in his recent report that the Work at Height Regulations need to be reviewed to ensure they do not lead people to go beyond what is proportionate or that the legislation was intended to cover.

I wonder if there should be a move to check they are being applied at all in some areas! The first enquiry was a call from a production manager who didn't understand the responsibility that he carried on behalf of the producers of a theatre show to operate safe systems of work. His lack of understanding about the forces generated by an arrested fall coupled with the desire for a cheap solution meant an awkward conversation about the absolute requirement not only to properly assess the risks to employees but also to select and provide appropriate work equipment.

If he allows an activity to take place, he has a duty to ensure it is as safe as possible under foreseeable circumstances. The injuries from a fall can be life threatening and therefore the control measures taken must be in proportion to that risk. Just because he has the genuine belief nobody was going to fall doesn't mean it can be ignored. If it is so 'safe', why use a fall arrest system? An employer has to supply work equipment to an appropriate standard, ensure it is used correctly and if PPE, to be compatible with the work, the user and other elements of the system in place. The technical crew had been issued new PPE and had training in its use but the lack of appropriate anchor points for their fall arrest PPE let the whole system down.

The second call was from another production manager, coincidentally also a safety advisor. His question concerned an employer of seasonal workers climbing on structures for outdoor events whose assumption 'they are riggers, they know what they are doing' was clearly insufficient. The 'riggers' in question were using wholly inappropriate parts of structures as fall arrest anchors, at heights where the lanyard length was unlikely to arrest the fall before the rigger hit the floor or something on the way down. Just because you can fit a 'scaff hook' on it doesn't mean it will work as an anchor point.

These workers are asked to show certificates of training for work at height before they are taken on and indeed often have certificates. It seems that often the course must have focused solely on the equipment and not prepared the trainee to ask about the structure provided to 'clip into'. So the employer, justifiably not an expert in all areas, has relied (let's assume) on a manager who has taken a training certificate as proof of competence. They need to know how to use the PPE and inspect it, but for pity's sake they also need to attach it to something that stands an earthly should the worst occur. That must be addressed by the employer or his representative at every site, and never become a generic 'have you got a harness and lanyard?' Tick box.

Simply sending people on a training course does not make them competent and certainly won't absolve the employer of his legal duties. The training alone or equipment by itself won't make the job safer. An employer needs to look at what knowledge is required for the activities being considered, assess the knowledge and skill required and ensure the skills of the workforce match those requirements. The training provided has to be adequate and even then the work should be supervised to ensure the planned system remains effective and safe. This means the trainer has to have experience of the type of work in question to ensure training is delivered in context. Someone who is trained to use fall arrest equipment in a sector where anchor bolts are routinely provided is unlikely to be competent to find or create anchor points on structures in our industry any more than we'd know how to access wind turbines.

The third enquiry was actually a whole string of e mails from a technician about access to the rigging points in a venue being suspect. 'Nobody is doing anything about it despite us telling the management'. Not an uncommon situation by any means. In this case a timber catwalk had originally a wooden handrail on one side only, but even this had allegedly been 'kicked off so people wouldn't rely on it' resulting in no handrails. A better solution might be to lock the door before someone falls to their death. In my opinion (and one I put immediately in writing) if the operator of the premises won't listen, the best solution would be to call the local authority. They could if necessary take enforcement action to prevent access until it had been repaired or more likely to have a good look at the whole venue as it seemed that the management may have been failing at several levels. (Pun intended).

For similar reasons to the employer in the previous examples, the venue operator didn't have the knowledge or experience to realise how dangerous the situation had become. They trust incoming contractors to be competent and don't have the wherewithal to check. Often contractors won't report shortcomings to protect their income and reputations. In this case, one young technician fortunately did take a longer term view and called the local authority so the situation could be resolved before anyone got hurt. However, as anyone with experience of our industry reading this will agree, it's about time employers and their managements understood it is they who are responsible for providing safe systems of work, not their employees…

SPARKS
SUSPEND
ON
STAGE

# Rigging Call #41

*May 2012*

### Do we need a rigger?

'We don't need a rigger, it's only four points'. There are plenty of managements who (though maybe not always consciously) are cutting corners because they are unaware of their responsibilities. If the producer or production manager ignores the risks or is unaware of their responsibilities, riggers often don't get employed. Sometimes it seems as though it is thought necessary to keep people in the dark in order to remain competitive. Empower your staff too much and they will start to put up costs. It can often take an accident for a producer or promoter to learn that having a competent person in charge of a lifting operation is a legal requirement.

Within the entertainment industry that competence was always difficult to define; but now there is at least a standard in the UK, a Level 3 National Rigging Certificate. There are competent riggers who do not hold the NRC qualification but gradually, as the scheme is better understood, most are getting qualified and the situation I'm describing will improve. It's important to realise that many people will be operating at a much higher level than that required at assessment, but it proves the rigger knows at least the basics. To my mind the most important being an understanding of their own responsibility and limitations.

The decisions about how to rig something, whenever they are made, are critical to safety to those on stage and to the public. Fingers generally only get pointed when something goes wrong or people get hurt. In law it is understood that an employer need not be an expert; the business owner is running a commercial enterprise. By the same token the law requires that employer to ensure competent people are used where necessary to ensure that people are protected from harm. The key words there are 'where necessary'. Hanging motors on existing house points and slinging a truss from them is admittedly not greatly demanding in terms of knowledge or skill, but the principles and responsibilities are no different to hanging a large system in an arena.

Many lighting technicians are perfectly competent to undertake basic rigging duties, just as there are many riggers competent to undertake lighting tasks, but both roles are necessary with regard to the responsibility an employer has. Now that there is an established measure of competence for riggers it seems a lot of people want to meet it, but perhaps without realising it is rather more than just 'the bus to work' for lighting or sound technicians. To demonstrate competence as a rigger, you need to be able to do more than hook a motor onto a beam clamp or build a ground support system. You wouldn't get a plumber to wire your house, so you don't ask an electrician to do your rigging. It's a proper job and it carries a responsibility on behalf of an employer. Few people object to waiting for the house electrician to tail in, it's expected. The rigging responsibility includes establishing if a building or structure can be used to support the lighting or sound system, scenery or other production requirement, how those elements are to be lifted and supported and then for carrying out the whole operation, including the de-rig, safely.

The bigger the show, the greater the competence required, but sufficient competence is still necessary for even the smallest of productions where rigging is involved. Another important part of this is the employer recognising that the competent person they are 'appointing' cannot be expected to take these rigging decisions on the day; they may need days or even weeks of planning, but too often we still hear 'in my day we just to make it up as we went along'. This is all about employers and senior management taking a chance based on their perception of the risks involved rather than the actual situation. How many catastrophic failures have there been owing to changes in the loads applied or last minute changes? A competent and responsible person with the authority to take rigging decisions unchallenged by those with a commercial interest is a fundamental requirement for all events.

Whilst it must be stated that training or a qualification will not stop accidents from happening, it will make them less likely. Most qualifications are broad in scope; it is rare they are tailored exactly to any one employer's particular needs. It's common to hear 'the rigging qualification doesn't cover what we do'. Likewise, 'we don't need to hire a rigger for something that simple' (meaning we don't foresee anything going wrong and that the cost is not something we can justify against the risk we foresee). It is viewed as a commercial risk – however if something does go wrong, the cost of not using competent staff may pale into insignificance when compared to the consequences of a serious accident.

If you think about the test required to obtain a driving licence in the UK, we have to demonstrate a range of skills and a depth of knowledge that few of us use on a day to day basis. The training for a 'MEWP licence' will rarely be applicable to our sector's needs (and it's not a licence) yet most employers and venues insist we all have one. The law requires an employer to provide adequate information and training, so if an employer feels the qualification isn't relevant to the skills their staff need but want confirmation that they are competent, they probably need to provide specific training with some form of in-house assessment. I doubt a rigging qualification will ever be a legal requirement, but every employer should realise that they have a legal responsibility to ensure that any rigging is carried out by competent people.

don't drop anything
don't drop anything
don't drop anything
don't drop anything
don't drop anything
don't drop anything
don't drop anything
don't drop any
RIG
OR
DIE

# Rigging Call #42

*July 2012*

### The Rules of Rigging

Rigging doesn't have many physical constants except gravity and its effects and the old joke:

> "There are two rules in rigging. Rule 1, don't drop anything.
> Rule 2, see Rule 1".

As riggers we are constantly fielding questions from people who want to avoid hiring riggers to do the job for them properly. The old chestnuts include which way up a shackle should be and if you are 'allowed' to put shackle pins in an O ring. The short answer is, like so many things in life – it depends. That's not the answer people want to hear. They want absolutes. Preferably cheap ones. "You can't put the pin of the shackle in the 'Spanset', it has to go in the O ring". Well, it would be better for the sling, but what about the shackles fighting each other in the O ring? Maybe on balance it would be better to suffer the sling getting a bit pinched in the shackle but keeping the metalwork from getting all twisted up? The usual decider in these decisions is the load carried. How heavy is it? If using two-tonne slings, 3.25 ton shackles and a three-tonne O ring to lift something that weighs 500kg it probably isn't going to do any damage. If the same assembly is used for a long time in one place I'd be more worried because of the combination of load and repetitive wear. In any case, I'd strongly recommend looking closely at the slings after the lift and destroying them if they showed any signs of damage.

No 'safety rule' says you have to use things twice. Of course, you can just over specify everything, but what happens when you run out of the heavy gear or something changes on site? Generally we have to get the job done; if we know our craft and understand the materials we're working with we'll be able to make informed decisions. We hear people say: "I was always taught that (insert method) is the best way to do that". Under what circumstances? Did you have the ideal selection of gear? How much time was available? All these things have a direct bearing on what happens during a job, and therefore I would also suggest to the experience of anyone who was there. You've seen something 'work', nothing broke, nobody got hurt and the show went up on time. But was it the right method, or just one of many that could have been used under those circumstances?

A production manager once said: "There's two ways of doing this job, there's your way and there's the right way" simply because my method was not one he'd seen before. Fair enough, it's his job, so we do it his way. But what if it was wrong? As I recall I was observing the principles of engineering as I understood them in slinging a truss, but clearly it didn't square with his experience. I dare say his method was equally effective, but it was different to mine. Therefore to him it was 'wrong'. If his method was wrong, he was insisting on the wrong method being used, possibly making the situation dangerous as a result and denying himself the chance of perhaps learning a more appropriate method. If his way was right and my way wrong, he would have improved the situation. If both methods were equally effective, we've all learned something from the experience.

We often learn by copying what we see to be effective – nothing went wrong. If we carry on copying 'mistakes' we don't progress. If we rely on blanket rules to keep us safe we don't learn the details of our craft. People accept blanket rules on grounds of 'safety' but then any variation from that comfort zone is 'dangerous'. If we don't see it going wrong or don't realise that but for the safety factors in the equipment it would have gone wrong, it is seen as 'safe'. We don't have X ray eyes so we can't see what's going on inside the materials we're using, how much tension there is, how close to the wind we are sailing. As one of my mentors used to say (a lot) 'rigging is not an exact science' and he was so right. Without using a load cell on every point it's impossible to know with any accuracy the load we're applying to any equipment or structure.

I have long believed that using load monitoring should be mandatory in most situations. To prove my point, just this week I was asked by colleagues to predict the tension in a hoist chain for a tower lifting operation we were carrying out.

Using a combination of nearly 40 years' experience and what I like to think of as near Brunelian** intuition, I managed to overestimate by something around 20%. Just goes to show. It could have been worse, I could have underestimated by 20% or I may have had no idea at all.

* Brunelian as in I.K. Brunel. I am not an alumnus of Brunel or any other University.

Under the circumstances it made little difference, it was nowhere near people and in fact we were using a load cell so we knew with certainty the loads we were dealing with and could work accordingly, even if that meant abandoning the lift. Very similar situations exist on shows going on right now all over the world, and some people working on some of those shows have no idea of the forces they are working with. Couple that with a lack of understanding about how materials behave and we have a real potential for disaster.

SAFETY IN CONTROL SYSTEMS
BASIC CONTROLLER
BICKEL-PICKLE
ETCP
Recognised Trainer
VPLT
NRC

# Rigging Call #43

*October 2012*

### The 2012 PLASA Rigging Conference

The 2012 PLASA Rigging Conference drew to a close on 11$^{th}$ September. The Conference Centre was had been buzzing with rigging talk from 9am on the 10$^{th}$. The opening speaker was Roy Bickel who spoke about his 50-year career to date. Roy started in working in circus in the early 60s and the early Disney shows in arenas and tents, through Broadway and rock and roll tours including Elvis Presley and many other household names. Roy is still working on projects to this day.

It was interesting to note that many of the methods of rigging we take for granted today were born of necessity – there were fewer limitations and it was all about putting shows on safely and quickly. Roy's colour code of red, white and blue for five, ten and 20 foot steels is still used all over the world. Interesting that the colours were chosen because being those of the 'Stars and Stripes' they were easily remembered. It was an inspiring speech and reminded us all why we do it. We can use technology to obtain data so we know the loads we are dealing with. Riggers should have the authority to manage dangerous situations that may threaten lives. Roy finished with a plea to work safely so we can all enjoy as long a career as he has.

Rinus Bakker and Roy Schilderman from ARGH (the Dutch association of riggers and grounders) talked about the significant differences in regulations that affect rigging between European countries. The minimum requirements laid down in the various Directives are frequently exceeded by many countries and this can lead to some difficulty when touring and challenge the expectations of a production when moving from one country to another. The idea of standardisation is not new, it was discussed last year, but it seemed that the general view was that is a good thing. I think next year we may get some way forward with agreeing at least on 'globally' acceptable methods of rigging.

Then followed a short presentation from Michelle Enright, and Lori Rubinstein, from PLASA about what a region interested in having a rigging certification scheme would need to think about if the PLASA model was to be used. Both the National Rigging Certificate and the

ETCP programme in the US require significant commitment in time and money in order to be effective. It is not an easy process and any prospective 'bar-raiser' should be aware that it is a long term process that can't be short-cut without losing sight of the reason for it in the first place.

Technical Director of Showquip NZ, Nick Barnfield, helped us understand better the dynamic loads in rigging applications. He started by defining some terms. The word 'dynamic' is often used when 'impact' or 'shock' (load) would be more correct. Nick had carried out some tests to establish how reliable the traditional 25% overload figure actually is. In fact many of the results he obtained were somewhat lower, but his advice was to continue using those figures as a conservative estimate.

Nick is representative of very few individuals who are both qualified engineers and hands-on riggers. When engineers like Nick make recommendations, you know you're dealing with someone who really understands what we do.

The last session on the first day was both amusing and factual. Robert Dean, who founded ZFX, gave us a history of the oldest effect in entertainment. It clearly showed how 'flying' had developed since Greek times to its renaissance in the late 1800s. Robert has a real philosophy on developing a flying effect and several times referred to the 'art' of flying which is nice to hear. Rigging is only a small part of what a flying company and its staff do. However, rigging can be a very creative job, something we forget if hauling motor chains all morning. For ZFX it is merely the bus to work.

Gert Jan Brouwer, MD of Frontline Rigging in Holland got 'Right to the Point' and explained to us his interest in improving the ergonomics in rigging grids. By designing beams that are accessible without over-reaching or crouching, points can be rigged and de-rigged more easily. Hauling through a sheave from below, as some of us know from experience, can have significant benefits to health as well as safety. Gert Jan's assertion that his teams can rig 200 motors in just a few hours is the proof.

An interesting point was raised; the hierarchy of control when applied to work at height typically requires the use of catwalks as the default starting position. However, it could be argued that rigging grid beams present fewer obstacles or obstructions to rig points and in combination with other improvements (such as steels with hooks instead of shackles) makes the work safer and more efficient than from a catwalk.

'Flip' Sandell from the HSE Entertainment and Leisure Section described what happens following an accident; proof if it were needed that you really have to be able to demonstrate you've done everything reasonable in preparation and on site to control how work is done, how people operate and what to do if things change. If everyone present communicates that message it will improve things for everyone. It really is so easy to go that tiny bit further and make sure everything nothing is left to chance.

Introduced by Chris Cronin, Malcolm Richards, a hugely experienced structural engineer in the entertainment industry, presented a session about the considerations in designing roof systems. Not surprisingly, the recent spate of collapses was mentioned and Scott Nachemann, author of the report on the Indiana State Fair collapse was on the panel together with structural engineers Abigail Matthews and Dragan Kuzmanov. Nick Barnfield chaired the discussion and Gavin Bye, head of the HSE's Construction Sector and Policy Unit, gave us the reasons why the HSE are taking over the enforcement of temporary demountable structures. Management of the site was uppermost in most peoples' minds. The structure itself may fail but by that time the site should have been evacuated to eliminate the possibility of loss of life. The person with authority to start this procedure is the key.

The 2012 Conference closed with Alan Jacobi, who opened the inaugural conference in 2010, chairing the now customary final question time session. It seemed to AJ and to many delegates that 'harmonisation' was probably the way forward; to agree common standards and methods wherever possible. This will take a bit of organising and PLASA could well be instrumental in doing so. By the time the next Conference comes round we will hopefully have some proposals and maybe even some answers.

CLAUDIO
ESPOSITO
=CE RIGGING=
CE
CE
CE
CE
CE
CE
CE MARKED:
CERTIFIED?

# Rigging Call #44

*January 2013*

## CE Marking

Our industry tends to borrow from other sectors or activities. We use all kinds of products that were never designed for how we use them. Many items such as long link chain, chain bags, the Tallescope and even scaffold tube have all been lifted from other industries because we find they work for us. We are a mainly resourceful bunch with enough common sense to know what's going to work but the idea about whether something is 'legal' or if there is some hidden reason why we shouldn't use something frequently seems a worry, maybe as a result of 'claim culture'.

Advice is frequently sought from outside the industry because we seem to think they might know better than us. Who knows the business better than us? My message is 'Do your homework and have courage!' It can lead to myths such as 'never connect steel and aluminium karabiners together' that I've heard aerial artistes recount, terrified they've got dangerous equipment. There is absolutely every reason to worry about galvanic corrosion between different metals left long-term in a damp atmosphere but not for aerial artistes occasionally hanging their silks for a matter of hours in rehearsal spaces or venues. As with many 'golden rules', the expert advice given was probably applicable at the time but taken out of context facts frequently get distorted at each re-telling.

Something we might be guilty of is misunderstanding what the regulations really seek to control. Treating components such as swivels or karabiners used with silks in aerial performance as lifting equipment, or even as PPE, is an irritating misconception. It is not lifting equipment even if a performer is actually lifted or lowered on it (see the Machinery (Safety) Regulations 2008) and it's certainly not PPE – but it is 'work equipment'. It does need inspecting by a competent person and the inspection should be recorded, though the information to be contained in the record is not specified.

By treating such items as lifting equipment or PPE you would be covering all bases, but often users of such items could do the inspections themselves after a bit of training rather than believing they have to outsource experts who issue 'LOLER certificates'. More correctly ‘accessories for lifting’, a lifting accessory is defined in law as ‘lifting equipment for attaching loads to machinery for lifting’. So slings and their components, shackles, rings and so on are accessories for lifting. It’s not as usual, but this could also include karabiners, maillons and similar specialist connectors. There is no reason that they can’t be used provided a risk assessment is carried out and the manufacturer’s instructions don’t forbid particular uses. A lifting accessory such as a shackle which is intended by the manufacturer to be used for lifting must (in Europe) bear a CE mark to signify conformity with the ‘essential health and safety requirements’ in the Machinery Directive before being placed on the market.

Products not designed for lifting should be confirmed as appropriate and fit for purpose by the employer (or self-employed person). That requires a bit of nous but doesn’t mean you can’t use them at all. We use many items of generic ‘lifting equipment’ which could be used this morning as a ‘lifting accessory’ but as lifting equipment (supporting the lifting machine, or ‘above the hook’) this afternoon. Long link ‘deck chain’ is a good example. When specified and used correctly it won’t make any difference where it is placed in the load path, provided the links are straight, not trapped or spread and only ever used in tension. Equipment used ‘below the hook’ in most industries is subjected to wear and tear and thus is required to be more robust and will require more frequent inspection.

In entertainment rigging we tend to use the same equipment in both applications and typically for temporary installations. As such we rig and de-rig frequently (albeit generally in a very clean environment in comparison with other industries), but we still treat what is technically lifting equipment, for example a beam-clamp, as though it were a lifting accessory. No bad thing given the risks in our sector; an example of how we actually do more than the law requires in certain areas. Selecting any item to use for lifting or suspending something means it has to be strong enough, but we are often in a situation where the compatibility between items in a load path is staggeringly out of proportion. Think of a typical chain hoist suspension. The weak link is probably the hoist itself, the slings and other accessories involved are frequently at least three times stronger. In a theatre environment, scenery may be hung from a ‘rated’ flying bar using drifts and flying irons which individually have a SWL equal to the total load but each may be connected to questionable timber scenery with a few woodscrews. We need to look at the bigger picture and ask what is the most likely thing to fail, and why. We check each sling, drift or shackle but frequently the method of connection to the scenery is ignored. In the case of the chain hoist, the truss supporting the load and the fittings connecting the load to the truss need to be considered. Even if the construction of the scenery is nothing to do with you, or the truss is slung by the production, if you're going to be lifting it you are partly responsible. If it falls apart whilst on the floor, it's not going to hurt anyone. If you agree to lift it and it disintegrates, guess whose responsibility it is then?

So in conclusion rather than worry about if something is 'legal' or not, apply the common sense principles that the law requires in any case. Every employer or self-employed person needs to select appropriate work equipment ensure those supervising, using and maintaining the equipment are trained and knowledgeable and that the equipment is inspected often enough to make sure it stays safe to use. As I seem to recall Harry Donovan saying, the best insurance is not to need any.

# Rigging Call #45

*March 2013*

## European Union?

This time, a report of an interesting two days with the European Arenas Association. (EAA) amongst riggers, rigging and technical managers and venue safety folk.

Formed in 1991, the EAA represents 31 arena venues spread across 20 countries. It is committed to promoting best practice amongst its members, and as part of that responsibility the Association's rigging guidance document has developed since a meeting in Stockholm in 2005. Following seminars held in Riga and Paris, the group convened again this January at the LG Arena (formerly the NEC Arena) near Birmingham in the UK.

Providing an opportunity for personnel with rigging responsibility to meet and share information, the seminar was chaired by Malcolm Close, director of operations at the Scottish Exhibition and Conference Centre. The event was superbly organised by Paul Rowlands and Ellie Coombes (rigging development manager and events manager for NEC Group Arenas respectively). 35 delegates comprising representatives from 16 EU arenas, four guest speakers and the EAA secretariat were welcomed to the LG Arena by Guy Dunstan, general manager, NEC Arenas Division.

Paul Rowlands inspired us with a short presentation on international co-operation, after which we toured the International Convention Centre and National Indoor Arena in Birmingham where Strictly Come Dancing had loaded in. Harry Box from UK Rigging spoke from long experience about how to approach efficient and safe rigging based on planning and communication. After lunch we returned to the LG Arena to hear a presentation from PLASA's Qualifications Manager, Michele Enright, about the international development of the National Rigging Certificate. There has been huge demand for a qualification internationally since the last EAA seminar in Riga and Michele explained how the scheme works and what is required in order for it to be effective using the 'NRC Nordic' recently launched in Scandinavia as an example.

Day two started with the NEC Group safety team's Sarah Green (aka 'Safety Sarah'), and chair of the National Arenas Association Safety

Advisory Group, explained the role of the SAG and Michael Anderson, Group head of health and safety, Earls Court and Olympia, talked about lessons learned from long experience of enforcing house safety policy. Is the job really so important to do so quickly that you can't do it safely? The time spent defending and correcting an unsafe method of work is frequently at least equal to the length of time it would have taken to do the job properly in the first place.

Then followed the serious business: reviewing the EAA 'Guidance for Rigging in European Arenas'. The document has existed for some time and contains the essential information required to support and advise rigging personnel. Time was spent working through roles and responsibilities in managing rigging including competence, legislation and accident reporting. More technical matters were also considered such as the specification of general PPE and that required for work at height. Discussions about inspection regimes led onto the more contentious area of working practices. The perennial issue of secondary suspensions arose and this led to an interesting discussion about when they are or are not necessary, focusing particularly on the difference between the UK and Germany.

In the UK we have choices; it would appear that in some parts of Germany at least, the requirement is for load to be taken off the hoists and applied to the secondaries, so the issue with slack or poorly installed 'safeties' doesn't occur as it does in the UK. We are increasingly standardising practices across Europe, though the secondary suspension issue demonstrates we are not out of the woods yet.

PLASA have commissioned a comparison of the rigging standards and codes of practice in existence in several countries to establish just where the common ground lies. Hopefully this will be reviewed at the 2013 PLASA Rigging Conference in October. The vision of an EU wide rigging code of practice is something that was almost mandated at the Conference last year, if only to agree the headings. Something as simple as agreeing the conventions used in a rigging plot would make life easier for venues and rigging contractors. This was strengthened further during the EAA seminar. It was unanimously agreed that every rigging plot should carry a key to the symbols used in it, legibly display each rigging point's X and Y coordinates, the load carried at each point in kilogrammes and the load capacity of each hoist detailed.

The new Hydro Arena is being built on the SECC site in Glasgow and due to open later in 2013. Malcolm Close gave a presentation on the background to its construction and a technical overview. This looks to be a very interesting building with a whopping rigging grid height of 37m. I feel sure it will be the subject of technical articles in the near future. Lastly we heard two presentations about mother-grids. Pedro Roberts, technical manager of the Globe Arena Stockholm presented a powerful argument for using a mother-grid as a rigging solution in a tall building. The grid can be travelled out to the centre of the arena if necessary and is lifted with 21 wire rope winches but without need for secondary suspensions. The type and configuration of the trusses in the Globe grid means that bridling between trusses is possible, not typical of mother-grids.

Adam Tullett, senior rigging coordinator at the NIA then spoke about the mother-grid at the NIA in Birmingham which operates a more traditional system using spreader trusses laid over the grid to provide the rigging points. This 30 x 25m grid can be travelled over the entire length of the arena, but is raised on 15 chain hoists and then deaded off at 5m centres thus optimising its load capacity. I was going to say it was a fitting close to the event because the NIA grid is Paul Rowlands' baby but at 21 years old, it is a baby no longer.

A very worthwhile event which is helping to standardise the industry for the benefit of all concerned and a precursor to the Riggers' Forum in Leeds at PLASA Focus in April, and the PLASA Rigging Conference at ExCel in October.

# Rigging Call #46

*May 2013*

### What does a Certificate Mean?

Several conversations over the last month or two have led me to ponder some of the effects of having a rigging certification scheme. One important question is often overlooked – what do non-riggers – promoters, producers, managements and technical or production managers, make of a rigging certificate?

'Rigging' means different things to different people. To some it may mean hanging chain motors, to others moving large and complex props or pieces of scenery or it may mean rigging banners and installing catenary wires for drapes in an exhibition hall. Rigging in the film and television sector meaning something slightly different again. For many freelance riggers it will be a mix of all of these things. A rigger requires a wide range of skills in most sectors of the industry and few of them can be learned overnight.

Why is it so often believed that having one rigger is enough? The work may need a rigging company, not just a rigger. Is this because employers really don't accept that you need sufficient competent people when it comes to rigging? A certification scheme may help with that. Even the word 'certificate' means different things to different people. Most people have their own criteria for what makes 'a rigger' but a qualification is based on attaining a standard and that will not match everyone's expectations. The rigger may have had relevant experience but it isn't a given simply because of 'having a certificate' – rigging's too big a subject.

We hear: 'How on earth did [that rigger] get his/her NRC?' Some people will scrape through; some will be capable of far more than is required but both have the qualification. It is ultimately up to an employer to determine if an individual is up to the mark and the certificate is a benchmark that assists that process. Back in the day there were significantly fewer shows and fewer people involved so the risks were statistically lower. We didn't want to pull the building down or damage people in roughly equal measure and were trusted not to do that. Now it seems nobody trusts anyone to act responsibly, usually as a result of someone's greed. You have to prove you're competent so someone can blame you if it goes wrong, but they also must take some responsibility for employing you in the first place; a qualification helps them do that. From a rigger's point of view it means that your skills are recognised and the qualification will support you when you have to stand your ground in maintaining safety standards.

Establishing a testing process forces an industry to examine itself in order to write the exam or design the skills test: asking why we do things a certain way, to demonstrate real understanding rather than ticking a box just because the candidate didn't drop anything. We all have personal beliefs and learned behaviours, some essential and some which are not always necessary. When one sits down to define the standards 'a competent rigger' should meet, it becomes very apparent that it can't be about opinion and it isn't just about rigging. A rigging qualification needs to reference best practice and is as much about the ability to communicate effectively and how to manage risk at work as it is about rigging.

This article is primarily about the National Rigging Certificate which is an assessment of competence – not a training course. If you work as a rigger and are competent in the way that the word is defined in legislation, you'll achieve the qualification. If you have occupational knowledge or skill gaps, you won't. The founding principles were that it should be of recognised quality, robust and transparent. It was put together for riggers by riggers and our largest trade association, informed by and shaped in line with accepted assessment practice against National Occupational Standards. We are now in a position where we have set the benchmarks, and as a result we have a qualification specific to entertainment rigging, not one from the construction industry forced upon us by the authorities.

It would be fantastic to have a degree level rigging qualification, but if £600 is a problem think what a degree would cost. Unfortunately it's ultimately a numbers game and there are just not enough riggers to justify the costs of offering a higher qualification, even if every venue insisted on only using degree level riggers. Hopefully before long every venue and licensing authority in the UK will demand the use of NRC qualified riggers to at least supervise, if not carry out, rigging work and that the scheme will develop to provide a safer working environment and a career path for future generations.

# Rigging Call #47

*August 2013*

## Whose gig is it anyway?

Inspired by various recent seminars and by the Riggers' Forum at PLASA Focus in Leeds, I've been thinking about the confusion that surrounds rigging roles and responsibilities. It seems to me to stem from two fundamental issues. The first is what I wrote about in the last Rigging Call – the need for venues and employers who are not rigging companies to understand what competence means in relation to rigging and thus what qualifications or certificates actually represent. The second is about knowing who is in control of whom.

The employment status of a freelance rigger and the relationship between that rigger and other people on site can be confusing with regard to who is in control. An employee of a rigging company is easier to define. You work for the rigging company directly, on PAYE or as a 'labour only sub-contractor'. You take direction from a manager or supervisor. If you are hired in as a freelance by a rigging company you are under the direction of the company that employs you. If you are hired by a crewing company, you'll be an employee of the crewing company (probably considered an 'employment business' in legal terms) but an employment business will often be supplying labour to work under the supervision of a client company.

There should be a contract between the crewing company and the promoter or producers or the production, usually via the production manager, with regard to who controls you. Depending on the nature of that contract, as a freelance rigger you will be working under the supervision of the crewing company, promoter/producers or the production. It is more widely understood these days that if you are a 'self-employed' person for income tax and National Insurance purposes, from a health and safety point of view in the examples given above you'd be treated as an employee.

Freelance riggers are 'labour only sub-contractors'. These are, for example, workers who supply their own tools but not the materials to do the job; workers who usually can't substitute other people to do the work as they might do if they were a company, and – crucially – they are people who don't take an entrepreneurial risk in the work being done. This is redolent of the 'lump labour' system in the UK construction industry during the 60s and 70s, whose workers' main concern was to finish the job quickly and move on to the next. That caused concerns over health and safety, the length of the working day and rates of pay.

Sound familiar? The fact of the matter is that usually the person who is legally responsible for you will be the person paying you unless a contractual arrangement exists between an employer and a client company to supply workers and that those workers are under the direction of that client company or its agents. To be a 'bona fide' self-employed sub-contractor usually means that you work for a number of clients, you've quoted for the work and you'd be liable for the financial consequences of any mistakes.

Many of the frustrations that riggers suffer are due to the fact that the people employing them don't understand rigging and who does what. An 'up-rigger'is frequently assumed to be the person with the greater skill, needing to be qualified and therefore the one that costs the money. The role of the 'down rigger' is an essential one but is frequently undervalued and often ignored completely. It's not OK to have a stagehand tying on motor chains, itself the cause of many accidents; it's a skilled job and a rigger should be employed to do it. Sending a qualified up-rigger and an unqualified down-rigger to a job 'because the down-rigger doesn't have to climb' is nonsense. The entire rigging operation is usually directed from the ground and the most experienced rigger will often be on the ground throughout because that's where decisions are taken and from where the most important communications emanate. Maybe the rigging process is just not understood? Maybe the experience of such employers is from a much earlier time?

Rarely should rigging be worked out 'on the day' any more. The rigging plot is 'advanced' by the tour rigger or master carpenter and the rigging is installed in accordance with that plot by riggers on site, whoever they work for. It may even be pre-rigged though inevitable adjustments take place on site, but the main structural and other essential decisions are typically taken months previously, especially for large scale and arena shows.

An up-rigger isn't necessarily the person who decides how much load a beam can take or even how the attachment is made; up-riggers in most cases are employed to make the attachment to the building, necessarily supported by down riggers. It's a team effort. The production rigging design work has been carried out by the head rigger or master carpenter on the production or by the rigging company who services the tour venues, or more likely, both. A venue provides the facility for the promoter to stage an event. The promoter pays the production. The production may employ up and down riggers via a crewing company. The riggers must be competent, including the down riggers. The venue, as the provider of the premises, has a duty to ensure that. The employer of the riggers, be it a rigging contractor, the production or a crewing company, needs to ensure that too.

For many of us, regulation isn't the reason we work in this industry, quite the opposite, it may turn a lot of us off; but we are always told health and safety is an essential part of good business and they do call it show *business*...

RIGGING PLOT
LIFTING PLAN
Lifting Operation
Scenic Movement
Risk Assessment
Weigth
Method

# Rigging Call #48

*September 2013*

## Competence

Like many people in the industry I attended the recent ABTT Theatre Show. A bustling, vibrant event now held in the Old Truman Brewery off Brick Lane in London. An extremely wide range of exhibitors from small drama schools to international lighting brands were represented in a very egalitarian way. By that I mean most exhibitors had a similar size stand in comparison with other trade shows. This has an interesting effect in as much as it almost forces people to mingle in the aisles. The ABTT show has always been a talking show, and it can take an hour to get a coffee simply because of bumping into people you can catch up with on the way.

This year the seminar programme tripled in size and covered a range of subjects dear to the hearts of the ABTT and they reflected the changing needs of the theatre sector and seemed very forward looking, preparing the next generation of technicians for employment in what has to be one of the most satisfying of careers.

I'm approaching the fiftieth piece I've done for Lighting+Sound International magazine and realise that although it's cropped up incidentally, competence is a subject I haven't covered. It was a word much in evidence at the Theatre Show. Ensuring competence in contractors, assessing competence and demonstrating competence are all everyday expressions. It's a word we're all familiar with but what is it exactly? It's something we all need to have when at work apparently, something desirable. Some people want to buy it in, some appreciate it takes a long time to acquire, others believe they already have enough and don't need any more.

The truth is that we all have an amount of competence in everything we do. To try to measure it without putting it in context is meaningless. I am expert at making a cup of (instant) coffee. I can be back at my desk without noticing I've made one. The point is I'm so practiced I know practically every possible pitfall. The task is to within a reasonable time scale make a cup of coffee that is acceptable to drink. To do this to a satisfactory standard means not being damaged by hot water, not wasting resources and not making a mess that needs cleaning up whilst still having something halfway drinkable at the end of the process. I can meet those requirements. The task is so straightforward and requires little or no planning to carry out. Provided there's coffee in the house, I have water, kettle, spoon and mug I'm sorted. You could say I'm competent – at making coffee the way I like it.

Now what if I try to make a cup in the dark? I draw on my experience; I predict what's likely to happen. I use the knowledge I have of boiling water, the location of the cold tap and where we keep the spoons. I assess the risk in lifting a hot kettle and gauging where the spout is relative to the mug. I'd manage. It would probably take ten times as long, make a mess and taste worse than usual, but I'd get there. But not as competently. So the point is that there is no such thing as wholly competent.

We each have a level of competence defined by the level of knowledge, amount of skill and the experience we have of something. That level will change every time we perform a task. It may increase every time we read about that subject or one that affects our thinking about it. We might go to a seminar and learn that a regulation has changed or even that one exists. A conversation with someone might increase our level of awareness or knowledge of a subject. So you could argue one never attains competence completely. Even because after being 50 years a flyman the theatre is refurbished with an automated flying system and you have to start all over again. The idea that competence can be acquired is a problem.

The current safety culture has to answer for some of this. We read regulations and codes of practice that require competent people to be employed. We are encouraged to look for a 'finished product' and we specify what that product needs to be according to what we are doing. I need a competent person to plan and supervise a lifting operation. What type of lifting operation? Bringing a bar in does not require a crane driver. It needs someone who has the appropriate knowledge, skill and experience to do the job safely, within time frame and using the resources available. That competence requires a particular blend of knowledge, skill and experience which is unique to that task.

It is usually the case that a range of competences is required to meet the requirements of a job description. Therefore to be a competent (insert job of choice) you need to be as competent *as necessary*, probably in a number of areas. Over-qualified in one area perhaps, just getting by in another; that's the situation for most of us. However, if we meet the criteria overall we are considered competent *enough* to carry out that function. It makes you realise how important job descriptions are and how much knowledge the management needs to have in order to structure an organisation. So when it comes to assessing competence, knowledge, skill and experience are equally important, in the proportions required for the task in hand. It is about finding out if someone knows what they are doing and why, if they can carry out the task in hand and if they can reliably take it in their stride. It's not about how much responsibility can be off-loaded onto them.

2013 PLASA
RIGGING CONFERENCE

# Rigging Call #49

*November 2013*

## The 4th PLASA Rigging Conference

The 4th PLASA Rigging Conference was held on 7th-8th October at London's ExCeL exhibition centre. This year sponsored by Unusual Rigging, Outback Rigging and Yutree Insurance, delegates from all over the world attended a total of eight sessions. The opening address for 2013 was given by Robin Elias, who epitomises the archetypical rigger. However, sex and drugs and rock and roll and experimentation with Tirfors and early trusses were balanced with a staggering list of achievements in suspending or manoeuvring objects such as planes and works of art. Robin will have inspired another generation of riggers to follow his lead in his pioneering approach. Robin often referred to the satisfaction of a job well done, often by himself and using his mental and physical strength to achieve the desired result – surely the reason we do it.

Cristiano Giavedoni is a regular participant in the conference. His innovative approach to presenting was displayed again, this time his supporting illustrations were detailed yet naïve cartoons depicting an imaginary voyage around four 'islands of safety'. Islands are frequently seen as attractive, but mysterious and elusive. Achieving a safe installation is a bit like discovering an island, but islands can be dreams not reality. The session demonstrated that to find the island, it is necessary to be able to identify its position in order to enjoy its benefits on subsequent visits.

Ron Bonner, PLASA's technical resources manager (EU) spoke about the upgrading of CWA 15902 to an EN and Paul Rowlands from the NEC Group Arenas announced the new European Arenas Association rigging code of practice. Although this may have seemed confusing in relation to the afternoon session, the new rigging guidance document doesn't reinvent the wheel, if anything it will act as a check method for future revision of such a document.

Lori Rubenstein and Michele Enright from PLASA had news of an initiative to create a template for an international rigging guidance document. The document will contain a comprehensive list of headings that should be included in any rigging code of practice. The features of a particular working practice will be supported by references to Standards and legal requirements in each of the three regions. It is believed that the document might be the framework document for rigging, a repository for standards and reference works, and a glossary of terms. This would inform employers, venue managers, training courses and enforcing authorities. The work continues and the intention is to have at least a draft document available by the 2014 conference.

Chris Bottger, tour rigger, trainer and author then gave us an overview and explanation of the features of the German BGV regulations. These are frequently thought of as onerous standards (though actually a code of practice) but they are not so different to the requirements of LOLER. This session sought to highlight the significant differences between the BGV and the rest of Europe. The bottom line is that in Germany there is no choice but to adhere to these regulations. They are designed to help minimise risk so insurance can be offered to all workers in Germany. It's not a bad idea – but needs to be harmonised with other European regulations.

HSE's Gavin Bull from the HSE (HM Inspector in the Construction Division) had the difficult job of firstly explaining the potential impact of the new CDM Regulations (2014) on the entertainment sector and secondly reporting that the consultation process is delayed. However, the discussion that followed was extremely interesting and covered topics including the critical role of 'the designer'.

An issue of concern was that the definition of construction is very wide and that even painting on site could mean that the work becomes notifiable under the CDM regulations 2014. Watch this space.

Malcolm Richards is a chartered structural engineer with many years' experience of designing aluminium structures. Malcolm's 2013 presentation was short and sweet but inspired many questions. The upshot of Malcolm's session was that even with finite element analysis programmes, a Brunellian approach makes us feel more comfortable; load a truss at a node point and try to get as much sling around as much metal as possible. Picking up a truss from the top chords makes a truss more stable, and picking up from the bottom chords ensures that the bracing is in compression not tension. Questions from the floor highlighted the fact that riggers are rarely engineers.

Scott Nacheman from Thornton Tomasetti delivered a very polished presentation about the lessons learned from the stage roof collapse at the Indiana State Fair in 2011. This tragedy was precipitated by the inadequacy of guy wire anchors. The resulting collapse caused seven deaths and it is clear there was no plan to evacuate the area before the structure was in danger of collapse. Many companies buy a 'roof system' but then reconfigure it to suit a particular project. If this reconfiguration deviates from the original design it should be re-evaluated by a competent person to ensure that it is appropriate for this new application. A reputable supplier will provide an engineering or structural report such that users' own engineers can provide this facility.

Mark Surtees, MD of Outback Rigging, summed up the conference and invited questions about programme topics. This included a discussion about the users of roof systems relying on the supplier's catalogue information rather than using the services of an engineer. An example of how the perceived cost of engineering reports can mean that structures get designed by people without the necessary competence. The industry needs reminding that it really must use engineers to design all but the simplest of bespoke structures; it is rarely a job for riggers.

All in all it was a worthwhile event to attend and one that should again get the full support of PLASA and the rigging industry on October 6, 2014.

CWA15902
CONVERSION WORK
DELAYED
BSI - DIN TC10 - CEN - EN - CWA
CWA 15902
EUROPEAN AGREEMENT

# Rigging Call #50

*January 2014*

**The times they are a-changin'…**

It is a time of change in the rigging world. There is a 'movement' trying to consolidate what we know and help riggers be a little more confident in their undoubted abilities. The National Rigging Advisory Group (NRAG) has recently facilitated this, as has ongoing work at PLASA, the ABTT, PSA and other bodies such as the Institute of Structural Engineers. Since September, NRAG has been strengthened and enlarged with representatives from most sectors of our industry and it seems to me that there are few divisions; we all want a safe and efficient way of getting shows on.

*This new 'movement' is starting to challenge accepted thinking about the safety of loads lifted overhead by chain hoists. Production equipment is usually either lifted and then left suspended overhead, or it is lifted but then moved overhead. This requires two different approaches. The CWA has several suggestions, but to cut to the chase, it is essential that the reliability of equipment used to lift or suspend loads overhead is never in question. That means doing a risk assessment and correctly specifying appropriate equipment and method of installation.*

Pay close attention also to the weight to be lifted. It seems to me that many 'accepted' safety measures in our industry have come about because accurate load information has been allowed to remain a secret. If you lift something without knowing its weight you are not working safely or legally, full stop. Load measurement is often essential for multiple suspensions or where unpredictable distribution of known loads is possible.

*Accepting sub-standard equipment or installation and then making it 'safe' to get the show on by using secondary suspensions is outdated, and to quote Technical Standards for Places of Entertainment, 'could promote a sense of false security'.*

If you want to suspend a load over people in the UK, you don't have to use C1 or Category A hoists because they have double brakes. *Specifying hoists with* double brakes to 'remove the need for safeties' is nonsense. Unless a failure affects the brake itself (a rare event), the number of brakes provided in a hoist is meaningless. Using the BGV C1 specification calls for far more than just double brakes; the user needs to observe the specification for the control system and a great many other factors to comply. Let's not forget that BGV hoist classifications are not Standards anyway, and standards are not usually legal requirements*. However it is a legal requirement if you intend to move a load overhead…in Germany.

If secondary suspensions are necessary, use them; but a risk assessment might suggest using a greater number of suspensions to support the load is better. Installing secondaries creates risk through additional working at height and also prevents the load from being moved in an emergency. Depending on circumstances and your risk assessment, you can use bog standard (Category B or D8) hoists and not necessarily even need to use 'safeties' if the load is supported sensibly. Generally the hoists in such cases should be loaded only to 50% of the manufacturer's rated capacity but even then, that comes from a code of practice, not a law.

Suppliers offer what they perceive the market wants. When users (and enforcers) are themselves misinformed or not competent the blind are leading the blind. This is why people think hoists are 'safer' if they meet this or that specification. The kit won't do the job for you.

With such potential and the progress already made it has come as a shock that the natural progression for CWA 15902 to become a full blown EN has been interrupted. Since 2008 CWA 15902 has been a voluntary agreement between trade associations, manufacturers, installers, consultants, users and public authorities within the entertainment industry across the EU. It's entitled 'Lifting and load-bearing equipment for stages and other production areas within the entertainment industry' and provides common specifications for national and international (travelling) performances, performers, technicians and manufacturers. It covers equipment that, as prescribed in European legislation, 'gives rise to danger'. The Machinery Directive and the Use of Work Equipment Directive are concerned with the minimum safety and health requirements of work equipment and state that appropriate procedures must be laid down and applied to protect workers that have to be present under suspended loads.

---

* I'm not suggesting we ignore standards necessarily, but have a look inside any British Standard and it says: Compliance with a British Standard does not of itself confer immunity from legal obligations. It will doubtless mitigate your case in a court, but is no 'Get out jail free card…'

The CWA considers these situations in in the entertainment industry, particularly when moving scenery or equipment over people, possibly in low light or with limited visibility during performances or rehearsals. CWA 15902 suggests how to maintain safety in these situations without making distinction between types of performance, venue or geographical location. To administer the transition from CEN Workshop Agreement (CWA) to a harmonised standard or EuroNorm (EN), interested parties must enlist the services of a Technical Committee (TC) at the Comité Européen de Normalisation (CEN). With no entertainment specific TC the needs of stage engineering was aligned to the nearest TC with similar engineering principles which was TC/10 lifts, escalators and moving walkways. A request was made to TC/10 by the entertainment industry to use TC/10 to promote CWA15902 to an EN standard. The application was heard at the end of November but a decision was placed on hold awaiting Germany's attempt to form an Entertainment TC at CEN. This application has been hurriedly put in place as Germany is alone in resisting CWA15902 as an EN. the EU might have to conform to a EU standard based on DIN 56950. This is fine if you are subject to the BGV, but not so helpful if you don't work in Germany.

# GLOSSARY OF TERMS IN THE CONTEXT USED IN THE ARTICLES

| | |
|---|---|
| **ABTT** | Association of British Theatre Technicians, founded 1960. |
| **ACoP** | Approved Code of Practice (not law but UK document with legal status) |
| **BGV** | Berufsgenossenschaft, statutory insurance for workers in Germany |
| **Category 3 PP** | The highest category, protecting the wearer from 'mortal danger' |
| **CWA** | Community Workshop Agreement – the next step down from an EN or Euronorm and really is a code of practice. CWA 15902 is entitled 'Lifting and load-bearing equipment for stages and other production areas within the entertainment industry. |
| **D8** | German classification (only) for standard industrial chain hoists |
| **Dynamic load** | Loads that are applied by motion, e.g. lifting a load against gravity |
| **Energy absorber** | Device to reduce the forces on the body in a fall arrest system |
| **EN361** | European standard for fall arrest harnesses |
| **EN362** | European standard for connectors used in a fall arrest system (karabiners, etc) |
| **ESTA** | North American trade association, now part of PLASA |
| **EWL** | Effective working length, e.g. the length of a roundsling as a loop or Ferrule Compression sleeve pressed around wire rope to form an eye termination |
| **HSE** | Health and Safety Executive (UK safety enforcement agency) |
| **Kader clamp** | Metal bracket fixing to extruded aluminium beams of modern marquees |

| | |
|---|---|
| **kN** | KiloNewton (unit of force rather than weight) 10kN = 1000kg force |
| **Lanyard** | Textile or rope element between harness and anchor PFPS |
| **LDI** | Lighting Dimensions International (trade show and magazine in the US) |
| **LOLER** | Lifting Operations and Lifting Equipment Regulations |
| **LSI magazine** | Lighting and Sound International (PLASA Media) |
| **LDI** | Lighting Dimensions International (trade show and magazine in the US) |
| **MEWP** | Mobile elevating work platform (e.g. cherry picker) |
| **NDT** | Non-destructuve testing |
| **Node or node point** | The 'strong point' in a structure where bracing has maximum effect |
| **O ring** | Steel oval ring used to form the apex or root of multi-legged slings |
| **Petzl** | French manufacturer of climbing, caving and industrial fall protection equipment |
| **PUWER** | Provision and Use of Work Equipment Regulations |
| **PFPS** | Personal fall prevention system |
| **PLASA** | Professional Lighting and Sound Association |
| **PSA** | Production Services Association |
| **Roundsling** | Endless sling comprising load-bearing yarn core and sheath (Spanset) |
| **Shear force** | Cross sectional 'chopping' force in a member, opposite of bending |
| **Socket** | Termination for wire rope using resin or molten metal to plug the rope end |
| **SWL** | Safe working load. The maximum load determined by the employer |
| **Syncope** | Fainting, passing out (pre-syncopal symptoms = those before passing out) |

| | |
|---|---|
| **Tallescope** | Wheeled vertical access machine; subject of much controversy in the UK |
| **Thimble** | Protective lining inside the eye of a fibre or wire rope termination |
| **Wire rope grips** | Clamps with threads used to form an eye termination in a wire rope |
| **WLL** | Working load limit: manufacturer's design load capacity of an item |
| **VLPT** | German trade association for Media and Event Technology |
| **Zarges** | Popular aluminium combination ladder |

## ENTERTAINMENT TECHNOLOGY PRESS

FREE SUBSCRIPTION SERVICE

Keeping Up To Date with

# 50 Rigging Calls

Entertainment Technology titles are continually up-dated, and all major changes and additions are listed in date order in the relevant dedicated area of the publisher's website. Simply go to the front page of www.etnow.com and click on the BOOKS button. From there you can locate the title and be connected through to the latest information and services related to the publication.

The author of the title welcomes comments and suggestions about the book and can be contacted by email at: cjhrig@btinternet.com

# Titles Published by Entertainment Technology Press

**50 Rigging Calls** *Chris Higgs, Cristiano Giavedoni 246pp* **£16.95**
ISBN: 9781904031758
Chris Higgs, author of ETP's two leading titles on rigging, An Introduction to Rigging in the Entertainment Industry and Rigging for Entertainment: Regulations and Practice, has collected together 50 articles he has provided regularly for Lighting + Sound International magazine from 2005 to date. They provide a wealth of information for those practising the craft within the entertainment technology industry. The book is profusely illustrated with caricature drawings by Christiano Giavedoni, featuring the popular rigging expert Mario.

**ABC of Theatre Jargon** *Francis Reid 106pp* **£9.95** ISBN: 9781904031093
This glossary of theatrical terminology explains the common words and phrases that are used in normal conversation between actors, directors, designers, technicians and managers.

**Aluminium Structures in the Entertainment Industry** *Peter Hind 234pp* **£24.95**
ISBN: 9781904031062
Aluminium Structures in the Entertainment Industry aims to educate the reader in all aspects of the design and safe usage of temporary and permanent aluminium structures specific to the entertainment industry – such as roof structures, PA towers, temporary staging, etc.

**AutoCAD – A Handbook for Theatre Users** *David Ripley 340pp* **£29.95**
ISBN: 9781904031741
From 'Setting Up' to 'Drawing in Three Dimensions' via 'Drawings Within Drawings', this compact and fully illustrated guide to AutoCAD covers everything from the basics to full colour rendering and remote 3D plotting. Third, completely revised edition, June 2014.

**Automation in the Entertainment Industry – A User's Guide** *Mark Ager and John Hastie 382pp* **£29.95** ISBN: 9781904031581
In the last 15 years, there has been a massive growth in the use of automation in entertainment, especially in theatres, and it is now recognised as its own discipline. However, it is still only used in around 5% of theatres worldwide. In the next 25 years, given current growth patterns, that figure will rise to 30%. This will mean that the majority of theatre personnel, including directors, designers, technical staff, actors and theatre management, will come into contact with automation for the first time at some point in their careers. This book is intended to provide insights and practical advice from those who use automation, to help the first-time user understand the issues and avoid the pitfalls in its implementation.

**Basics – A Beginner's Guide to Lighting Design** *Peter Coleman 92pp* **£9.95**
ISBN: 9781904031413
The fourth in the author's 'Basics' series, this title covers the subject area in four main sections: The Concept, Practical Matters, Related Issues and The Design Into Practice. In an

area that is difficult to be definitive, there are several things that cross all the boundaries of all lighting design and it's these areas that the author seeks to help with.

**Basics – A Beginner's Guide to Special Effects** *Peter Coleman 82pp* **£9.95**
ISBN: 9781904031338
This title introduces newcomers to the world of special effects. It describes all types of special effects including pyrotechnic, smoke and lighting effects, projections, noise machines, etc. It places emphasis on the safe storage, handling and use of pyrotechnics.

**Basics – A Beginner's Guide to Stage Lighting** *Peter Coleman 86pp* **£9.95**
ISBN: 9781904031208
This title does what it says: it introduces newcomers to the world of stage lighting. It will not teach the reader the art of lighting design, but will teach beginners much about the 'nuts and bolts' of stage lighting.

**Basics – A Beginner's Guide to Stage Sound** *Peter Coleman 86pp* **£9.95**
ISBN: 9781904031277
This title does what it says: it introduces newcomers to the world of stage sound. It will not teach the reader the art of sound design, but will teach beginners much about the background to sound reproduction in a theatrical environment.

**Basics: A Beginner's Guide to Stage Management** *Peter Coleman 64pp* **£7.95**
ISBN: 9781904031475
The fifth in Peter Coleman's popular 'Basics' series, this title provides a practical insight into, and the definition of, the role of stage management. Further chapters describe Cueing or 'Calling' the Show (the Prompt Book), and the Hardware and Training for Stage Management. This is a book about people and systems, without which most of the technical equipment used by others in the performance workplace couldn't function.

**Building Better Theaters** *Michael Mell 180pp* **£16.95** ISBN: 9781904031406
A title within our Consultancy Series, this book describes the process of designing a theatre, from the initial decision to build through to opening night. Michael Mell's book provides a step-by-step guide to the design and construction of performing arts facilities. Chapters discuss: assembling your team, selecting an architect, different construction methods, the architectural design process, construction of the theatre, theatrical systems and equipment, the stage, backstage, the auditorium, ADA requirements and the lobby. Each chapter clearly describes what to expect and how to avoid surprises. It is a must-read for architects, planners, performing arts groups, educators and anyone who may be considering building or renovating a theatre.

**Carry on Fading** *Francis Reid 216pp* **£20.00** ISBN: 9781904031642
This is a record of five of the best years of the author's life. Years so good that the only downside is the pangs of guilt at enjoying such contentment in a world full of misery induced by greed, envy and imposed ideologies. Fortunately Francis' DNA is high on luck, optimism and blessing counting.

**Case Studies in Crowd Management**
*Chris Kemp, Iain Hill, Mick Upton, Mark Hamilton 206pp* **£16.95**
ISBN: 9781904031482
This important work has been compiled from a series of research projects carried out by the staff of the Centre for Crowd Management and Security Studies at Buckinghamshire Chilterns University College (now Bucks New University), and seminar work carried out in Berlin and Groningen with partner Yourope. It includes case studies, reports and a crowd management safety plan for a major outdoor rock concert, safe management of rock concerts utilising a triple barrier safety system and pan-European Health & Safety Issues.

**Case Studies in Crowd Management, Security and Business Continuity**
*Chris Kemp, Patrick Smith 274pp* **£24.95** ISBN: 9781904031635
The creation of good case studies to support work in progress and to give answers to those seeking guidance in their quest to come to terms with perennial questions is no easy task. The first Case Studies in Crowd Management book focused mainly on a series of festivals and events that had a number of issues which required solving. This book focuses on a series of events that had major issues that impacted on the every day delivery of the events researched.

**Close Protection – The Softer Skills** *Geoffrey Padgham 132pp* **£11.95**
ISBN: 9781904031390
This is the first educational book in a new 'Security Series' for Entertainment Technology Press, and it coincides with the launch of the new 'Protective Security Management' Foundation Degree at Buckinghamshire Chilterns University College (now Bucks New University). The author is a former full-career Metropolitan Police Inspector from New Scotland Yard with 27 years' experience of close protection (CP). For 22 of those years he specialised in operations and senior management duties with the Royalty Protection Department at Buckingham Palace, followed by five years in the private security industry specialising in CP training design and delivery. His wealth of protection experience comes across throughout the text, which incorporates sound advice and exceptional practical guidance, subtly separating fact from fiction. This publication is an excellent form of reference material for experienced operatives, students and trainees.

**A Comparative Study of Crowd Behaviour at Two Major Music Events**
*Chris Kemp, Iain Hill, Mick Upton 78pp* **£7.95** ISBN: 9781904031253
A compilation of the findings of reports made at two major live music concerts, and in particular crowd behaviour, which is followed from ingress to egress.

**Control Freak** *Wayne Howell 270pp* **£28.95** ISBN: 9781904031550
Control Freak is the second book by Wayne Howell. It provides an in depth study of DMX512 and the new RDM (Remote Device Management) standards. The book is aimed at both users and developers and provides a wealth of real world information based on the author's twenty year experience of lighting control.

**Copenhagen Opera House** *Richard Brett and John Offord 272pp* **£32.00**
ISBN: 9781904031420
Completed in a little over three years, the Copenhagen Opera House opened with a royal gala performance on 15th January 2005. Built on a spacious brown-field site, the building is a landmark venue and this book provides the complete technical background story to an opera house set to become a benchmark for future design and planning. Sixteen chapters by relevant experts involved with the project cover everything from the planning of the auditorium and studio stage, the stage engineering, stage lighting and control and architectural lighting through to acoustic design and sound technology plus technical summaries.

**Cue 80** *Francis Reid 310pp* **£17.95** ISBN: 9781904031659
Although Francis Reid's work in theatre has been visual rather than verbal, writing has provided crucial support. Putting words on paper has been the way in which he organised and clarified his thoughts. And in his self-confessed absence of drawing skills, writing has helped him find words to communicate his visual thinking in discussions with the rest of the creative team. As a by-product, this process of searching for the right words to help formulate and analyse ideas has resulted in half-a-century of articles in theatre journals. Cue 80 is an anthology of these articles and is released in celebration of Francis' 80th birthday.

**The DMX 512-A Handbook – Design and Implementation of DMX Enabled Products and Networks** *James Eade 150pp* **£13.95** ISBN: 9781904031727
This guidebook was originally conceived as a guide to the new DMX512-A standard on behalf of the ESTA Controls Protocols Working Group (CPWG). It has subsequently been updated and is aimed at all levels of reader from technicians working with or servicing equipment in the field as well as manufacturers looking to build in DMX control to their lighting products. It also gives thorough guidance to consultants and designers looking to design DMX networks.

**Electric Shadows: an Introduction to Video and Projection on Stage** *Nick Moran 234pp* **£23.95** ISBN: 9781904031734
Electric Shadows aims to guide the emerging video designer through the many simple and difficult technical and aesthetic choices and decisions he or she has to make in taking their design from outline idea through to realisation. The main body of the book takes the reader through the process of deciding what content will be projected onto what screen or screens to make the best overall production design. The book will help you make electric shadows that capture the attention of your audience, to help you tell your stories in just the way you want.

**Electrical Safety for Live Events** *Marco van Beek 98pp* **£16.95** ISBN: 9781904031284
This title covers electrical safety regulations and good practise pertinent to the entertainment industries and includes some basic electrical theory as well as clarifying the "do's and don't's" of working with electricity.

**Entertainment in Production Volume 1: 1994-1999** *Rob Halliday 254pp* **£24.95**
ISBN: 9781904031512

**Entertainment in Production Volume 2: 2000-2006** *Rob Halliday 242poo* £24.95
ISBN: 9781904031529
Rob Halliday has a dual career as a lighting designer/programmer and author and in these two volumes he provides the intriguing but comprehensive technical background stories behind the major musical productions and other notable projects spanning the period 1994 to 2005. Having been closely involved with the majority of the events described, the author is able to present a first-hand and all-encompassing portrayal of how many of the major shows across the past decade came into being. From *Oliver!* and *Miss Saigon* to *Mamma Mia!* and *Mary Poppins*, here the complete technical story unfolds. The books, which are profusely illustrated, are in large part an adapted selection of articles that first appeared in the magazine *Lighting&Sound International*.

**Entertainment Technology Yearbook 2008** *John Offord 220pp* **£14.95**
ISBN: 9781904031543
The Entertainment Technology Yearbook 2008 covers the year 2007 and includes picture coverage of major industry exhibitions in Europe compiled from the pages of Entertainment Technology magazine and the etnow.com website, plus articles and pictures of production, equipment and project highlights of the year.

**The Exeter Theatre Fire** *David Anderson 202pp* **£24.95** ISBN: 9781904031130
This title is a fascinating insight into the events that led up to the disaster at the Theatre Royal, Exeter, on the night of September 5th 1887. The book details what went wrong, and the lessons that were learned from the event.

**Fading Light – A Year in Retirement** *Francis Reid 134pp* **£14.95**
ISBN: 9781904031352
Francis Reid, the lighting industry's favourite author, describes a full year in retirement. "Old age is much more fun than I expected," he says. Fading Light describes visits and experiences to the author's favourite theatres and opera houses, places of relaxation and re-visits to scholarly institutions.

**Focus on Lighting Technology** *Richard Cadena 120pp* **£17.95** ISBN: 9781904031147
This concise work unravels the mechanics behind modern performance lighting and appeals to designers and technicians alike. Packed with clear, easy-to-read diagrams, the book provides excellent explanations behind the technology of performance lighting.

**The Followspot Guide** *Nick Mobsby 450pp* **£28.95** ISBN: 9781904031499
The first in ETP's Equipment Series, Nick Mobsby's Followspot Guide tells you everything you need to know about followspots, from their history through to maintenance and usage. Its pages include a technical specification of 193 followspots from historical to the latest versions from major manufacturers.

**From Ancient Rome to Rock 'n' Roll – a Review of the UK Leisure Security Industry**
*Mick Upton 198pp* **£14.95** ISBN: 9781904031505
From stewarding, close protection and crowd management through to his engagement as a senior consultant Mick Upton has been ever present in the events industry. A founder of ShowSec International in 1982 he was its chairman until 2000. The author has led the way on training within the sector. He set up the ShowSec Training Centre and has acted as a consultant at the Bramshill Police College. He has been prominent in the development of courses at Buckinghamshire New University where he was awarded a Doctorate in 2005. Mick has received numerous industry awards. His book is a personal account of the development and professionalism of the sector across the past 50 years.

**Gobos for Image Projection** *Michael Hall and Julie Harper 176pp* **£25.95**
ISBN: 9781904031628
In this first published book dedicated totally to the gobo, the authors take the reader through from the history of projection to the development of the present day gobo. And there is broad practical advice and ample reference information to back it up. A feature of the work is the inclusion, interspersed throughout the text, of comment and personal experience in the use and application of gobos from over 25 leading lighting designers worldwide.

**Health and Safety Aspects in the Live Music Industry** *Chris Kemp, Iain Hill 300pp*
**£30.00** ISBN: 9781904031222
This major work includes chapters on various safety aspects of live event production and is written by specialists in their particular areas of expertise.

**Health and Safety Management in the Live Music and Events Industry** *Chris Hannam 480pp* **£25.95** ISBN: 9781904031307
This title covers the health and safety regulations and their application regarding all aspects of staging live entertainment events, and is an invaluable manual for production managers and event organisers.

**Hearing the Light – 50 Years Backstage** *Francis Reid 280pp* **£24.95**
ISBN: 9781904031185
This highly enjoyable memoir delves deeply into the theatricality of the industry. The author's almost fanatical interest in opera, his formative period as lighting designer at Glyndebourne and his experiences as a theatre administrator, writer and teacher make for a broad and unique background.

**An Introduction to Rigging in the Entertainment Industry** *Chris Higgs 272pp* **£24.95**
ISBN: 9781904031123
This title is a practical guide to rigging techniques and practices and also thoroughly covers safety issues and discusses the implications of working within recommended guidelines and regulations. Second edition revised September 2008.

**Let There be Light – Entertainment Lighting Software Pioneers in Conversation**
*Robert Bell 390pp* **£32.00** ISBN: 9781904031246
Robert Bell interviews a distinguished group of software engineers working on entertainment lighting ideas and products.

**Light and Colour Filters** *Michael Hall and Eddie Ruffell 286pp* **£23.95**
ISBN: 9781904031598
Written by two acknowledged and respected experts in the field, this book is destined to become the standard reference work on the subject. The title chronicles the development and use of colour filters and also describes how colour is perceived and how filters function. Up-to-date reference tables will help the practitioner make better and more specific choices of colour.

**Lighting for Roméo and Juliette** *John Offord 172pp* **£26.95** ISBN: 9781904031161
John Offord describes the making of the Vienna State Opera production from the lighting designer's viewpoint – from the point where director Jürgen Flimm made his decision not to use scenery or sets and simply employ the expertise of lighting designer Patrick Woodroffe.

**Lighting Systems for TV Studios** *Nick Mobsby 570pp* **£45.00** ISBN: 9781904031000
*Lighting Systems for TV Studios*, now in its second edition, is the first book specifically written on the subject and has become the 'standard' resource work for studio planning and design covering the key elements of system design, luminaires, dimming, control, data networks and suspension systems as well as detailing the infrastructure items such as cyclorama, electrical and ventilation. TV lighting principles are explained and some history on TV broadcasting, camera technology and the equipment is provided to help set the scene! The second edition includes applications for sine wave and distributed dimming, moving lights, Ethernet and new cool lamp technology.

**Lighting Techniques for Theatre-in-the-Round** *Jackie Staines 188pp* **£24.95**
ISBN: 9781904031017
Lighting Techniques for Theatre-in-the-Round is a unique reference source for those working on lighting design for theatre-in-the-round for the first time. It is the first title to be published specifically on the subject and it also provides some anecdotes and ideas for more challenging shows, and attempts to blow away some of the myths surrounding lighting in this format.

**Lighting the Diamond Jubilee Concert** *Durham Marenghi 102pp* **£19.95**
ISBN: 9781904031673
In this highly personal landmark document the show's lighting designer Durham Marenghi pays tribute to the team of industry experts who each played an important role in bringing the Diamond Jubilee Concert to fruition, both for television and live audiences. The book contains colour production photography throughout and describes the production processes and the thinking behind them. In his Foreword, BBC Executive Producer Guy Freeman states: "Working with the whole lighting team on such a special project was a real treat for me and a fantastic achievement for them, which the pages of this book give a remarkable insight into."

**Lighting the Stage** *Francis Reid 120pp* **£14.95** ISBN: 9781904031086
Lighting the Stage discusses the human relationships involved in lighting design – both between people, and between these people and technology. The book is written from a highly personal viewpoint and its 'thinking aloud' approach is one that Francis Reid has used in his writings over the past 30 years.

**Miscellany of Lighting and Stagecraft** *Michael Hall & Julie Harper 222pp* **£22.95**
ISBN: 9781904031680
This title will help schools, colleges, amateurs, technicians and all those interested in practical theatre and performance to understand, in an entertaining and informative way, the key backstage skills. Within its pages, numerous professionals share their own special knowledge and expertise, interspersed with diversions of historic interest and anecdotes from those practising at the front line of the industry. As a result, much of the advice and skills set out have not previously been set in print. The editors' intention with this book is to provide a Miscellany that is not ordered or categorised in strict fashion, but rather encourages the reader to flick through or dip into it, finding nuggets of information and anecdotes to entertain, inspire and engender curiosity – also to invite further research or exploration and generally encourage people to enter the industry and find out for themselves.

**Mr Phipps' Theatre** *Mark Jones, John Pick 172pp* £17.95 ISBN: 9781904031383
Mark Jones and John Pick describe "The Sensational Story of Eastbourne's Royal Hippodrome" – formerly Eastbourne Theatre Royal. An intriguing narrative, the book sets the story against a unique social history of the town. Peter Longman, former director of The Theatres Trust, provides the Foreword.

**Northen Lights** *Michael Northen 256pp* **£17.95** ISBN: 9781904031666
Many books have been written by famous personalities in the theatre about their lives and work. However this is probably one of the first memoirs by someone who has spent his entire career behind scenes, and not in front of the footlights. As a lighting designer and as consultant to designers and directors, Michael Northen worked through an exciting period of fifty years of theatrical history from the late nineteen thirties in theatres in the UK and abroad, and on productions ranging from Shakespeare, opera and ballet to straight plays, pantomimes and cabaret. This is not a complicated technical text book, but is intended to give an insight into some of the 300 productions in which he had been involved and some of the directors, the designers and backstage staff he have worked with, viewed from a new angle.

**Pages From Stages** *Anthony Field 204pp* **£17.95** ISBN: 9781904031260
Anthony Field explores the changing style of theatres including interior design, exterior design, ticket and seat prices, and levels of service, while questioning whether the theatre still exists as a place of entertainment for regular theatre-goers.

**Performing Arts Technical Training Handbook 2013/2014** *ed: John Offord 304pp*
**£19.95** ISBN: 9781904031710
Published in association with the ABTT (Association of British Theatre Technicians), this important Handbook, now in its third edition, includes fully detailed and indexed entries

describing courses on backstage crafts offered by over 100 universities and colleges across the UK. A completely new research project, with accompanying website, the title also includes articles with advice for those considering a career 'behind the scenes', together with contact information and descriptions of the major organisations involved with industry training – plus details of companies offering training within their own premises.

**Practical Dimming** *Nick Mobsby 364pp* **£22.95** ISBN: 97819040313444
This important and easy to read title covers the history of electrical and electronic dimming, how dimmers work, current dimmer types from around the world, planning of a dimming system, looking at new sine wave dimming technology and distributed dimming. Integration of dimming into different performance venues as well as the necessary supporting electrical systems are fully detailed. Significant levels of information are provided on the many different forms and costs of potential solutions as well as how to plan specific solutions. Architectural dimming for the likes of hotels, museums and shopping centres is included. Practical Dimming is a companion book to Practical DMX and is designed for all involved in the use, operation and design of dimming systems.

**Practical DMX** *Nick Mobsby 276pp* **£16.95** ISBN: 9781904031369
In this highly topical and important title the author details the principles of DMX, how to plan a network, how to choose equipment and cables, with data on products from around the world, and how to install DMX networks for shows and on a permanently installed basis. The easy style of the book and the helpful fault finding tips, together with a review of different DMX testing devices provide an ideal companion for all lighting technicians and system designers. An introduction to Ethernet and Canbus networks are provided as well as tips on analogue networks and protocol conversion. It also includes a chapter on Remote Device Management.

**A Practical Guide to Health and Safety in the Entertainment Industry**
*Marco van Beek 120pp* **£14.95** ISBN: 9781904031048
This book is designed to provide a practical approach to Health and Safety within the Live Entertainment and Event industry. It gives industry-pertinent examples, and seeks to break down the myths surrounding Health and Safety.

**Production Management** *Joe Aveline 134pp* **£17.95** ISBN: 9781904031109
Joe Aveline's book is an in-depth guide to the role of the Production Manager, and includes real-life practical examples and 'Aveline's Fables' – anecdotes of his experiences with real messages behind them.

**Rigging for Entertainment: Regulations and Practice** *Chris Higgs 156pp* **£19.95**
ISBN: 9781904031215
Continuing where he left off with his highly successful An Introduction to Rigging in the Entertainment Industry, Chris Higgs' second title covers the regulations and use of equipment in greater detail.

**Rock Solid Ethernet** *Wayne Howell 304pp* **£23.95** ISBN: 9781904031697
Now in its third completely revised and reset edition, Rock Solid Ethernet is aimed specifically at specifiers, installers and users of entertainment industry systems, and will give the reader a thorough grounding in all aspects of computer networks, whatever industry they may work in. The inclusion of historical and technical 'sidebars' make for an enjoyable as well as an informative read.

**Sixty Years of Light Work** *Fred Bentham 450pp* **£26.95** ISBN: 9781904031079
This title is an autobiography of one of the great names behind the development of modern stage lighting equipment and techniques. It includes a complete facsimile of the famous Strand Electric Catalogue of May 1936 – a reference work in itself.

**Sound for the Stage** *Patrick Finelli 218pp* **£24.95** ISBN: 9781904031154
Patrick Finelli's thorough manual covering all aspects of live and recorded sound for performance is a complete training course for anyone interested in working in the field of stage sound, and is a must for any student of sound.

**Stage Automation** *Anton Woodward 128pp* **£12.95** ISBN: 9781904031567
The purpose of this book is to explain the stage automation techniques used in modern theatre to achieve some of the spectacular visual effects seen in recent years. The book is targeted at automation operators, production managers, theatre technicians, stage engineering machinery manufacturers and theatre engineering students. Topics are covered in sufficient detail to provide an insight into the thought processes that the stage automation engineer has to consider when designing a control system to control stage machinery in a modern theatre. The author has worked on many stage automation projects and developed the award-winning Impressario stage automation system.

**Stage Lighting Design in Britain: The Emergence of the Lighting Designer, 1881-1950**
*Nigel Morgan 300pp* **£17.95** ISBN: 9781904031345
This title sets out to ascertain the main course of events and the controlling factors that determined the emergence of the theatre lighting designer in Britain, starting with the introduction of incandescent electric light to the stage, and ending at the time of the first public lighting design credits around 1950. The book explores the practitioners, equipment, installations and techniques of lighting design.

**Stage Lighting for Theatre Designers** *Nigel Morgan 124pp* **£17.95**
ISBN: 9781904031192
This is an updated second edition of Nigel Morgan's popular book for students of theatre design – outlining all the techniques of stage lighting design.

**Technical Marketing Techniques** *David Brooks, Andy Collier, Steve Norman 160pp*
**£24.95** ISBN: 9781904031031
Technical Marketing is a novel concept, defined and elaborated by the authors of this book, with business-to-business companies competing in fast developing technical product sectors.

**Technical Standards for Places of Entertainment** *ABTT 354pp A4* **£45.00**
ISBN: 9781904031703
Technical Standards for Places of Entertainment details the necessary physical standards required for entertainment venues. Known in the industry as the "Yellow Book" the latest completely revised edition was first published in June 2013.

**Theatre Engineering and Stage Machinery** *Toshiro Ogawa 332pp* **£30.00**
ISBN: 9781904031024
Theatre Engineering and Stage Machinery is a unique reference work covering every aspect of theatrical machinery and stage technology in global terms, and across the complete historical spectrum. Revised February 2007.

**Theatre Lighting in the Age of Gas** *Terence Rees 232pp* **£24.95**
ISBN: 9781904031178
Entertainment Technology Press has republished this valuable historic work previously produced by the Society for Theatre Research in 1978. Theatre Lighting in the Age of Gas investigates the technological and artistic achievements of theatre lighting engineers from the 1700s to the late Victorian period.

**Theatre Space: A Rediscovery Reported** *Francis Reid 238pp* **£19.95**
ISBN: 9781904031437
In the post-war world of the 1950s and 60s, the format of theatre space became a matter for a debate that aroused passions of an intensity unknown before or since. The proscenium arch was clearly identified as the enemy, accused of forming a barrier to disrupt the relations between the actor and audience. An uneasy fellow-traveller at the time, Francis Reid later recorded his impressions whilst enjoying performances or working in theatres old and new and this book is an important collection of his writings in various theatrical journals from 1969-2001 including his contribution to the Cambridge Guide to the Theatre in 1988. It reports some of the flavour of the period when theatre architecture was rediscovering its past in a search to establish its future.

**Theatres of Achievement** *John Higgins 302pp* **£29.95** ISBN: 9781904031376
John Higgins affectionately describes the history of 40 distinguished UK theatres in a personal tribute, each uniquely illustrated by the author. Completing each profile is colour photography by Adrian Eggleston.

**Theatric Tourist** *Francis Reid 220pp* **£19.95** ISBN: 9781904031468
*Theatric Tourist* is the delightful story of Francis Reid's visits across more than 50 years to theatres, theatre museums, performances and even movie theme parks. In his inimitable style, the author involves the reader within a personal experience of venues from the Legacy of Rome to theatres of the Renaissance and Eighteenth Century Baroque and the Gustavian Theatres of Stockholm. His performance experiences include Wagner in Beyreuth, the Pleasures of Tivoli and Wayang in Singapore. This is a 'must have' title for those who are as "incurably stagestruck" as the author.

**Through the Viewfinder** *Jeremy Hoare 276pp* **£21.95** ISBN:: 9781904031574
Do you want to be a top television cameraman? Well this is going to help!
Through the Viewfinder is aimed at media students wanting to be top professional television cameramen – but it will also be of interest to anyone who wants to know what goes on behind the cameras that bring so much into our homes.
The author takes his own opinionated look at how to operate a television camera based on 23 years' experience looking through many viewfinders for a major ITV network company. Based on interviews with people he has worked with, all leaders in the profession, the book is based on their views and opinions and is a highly revealing portrait of what happens behind the scenes in television production from a cameraman's point of view.

**Walt Disney Concert Hall – The Backstage Story** *Patricia MacKay & Richard Pilbrow 250pp* **£28.95** ISBN: 9781904031239
Spanning the 16-year history of the design and construction of the Walt Disney Concert Hall, this book provides a fresh and detailed behind the scenes story of the design and technology from a variety of viewpoints. This is the first book to reveal the "process" of the design of a concert hall.

**Yesterday's Lights – A Revolution Reported** *Francis Reid 352pp* **£26.95**
ISBN: 9781904031321
Set to help new generations to be aware of where the art and science of theatre lighting is coming from – and stimulate a nostalgia trip for those who lived through the period, Francis Reid's latest book has over 350 pages dedicated to the task, covering the 'revolution' from the fifties through to the present day. Although this is a highly personal account of the development of lighting design and technology and he admits that there are 'gaps', you'd be hard put to find anything of significance missing.

*Go to www.etbooks.co.uk for full details of above titles and secure online ordering facilities. Most books also available for Kindle.*